BestMasters

Mit „**BestMasters**" zeichnet Springer die besten Masterarbeiten aus, die an renommierten Hochschulen in Deutschland, Österreich und der Schweiz entstanden sind. Die mit Höchstnote ausgezeichneten Arbeiten wurden durch Gutachter zur Veröffentlichung empfohlen und behandeln aktuelle Themen aus unterschiedlichen Fachgebieten der Naturwissenschaften, Psychologie, Technik und Wirtschaftswissenschaften. Die Reihe wendet sich an Praktiker und Wissenschaftler gleichermaßen und soll insbesondere auch Nachwuchswissenschaftlern Orientierung geben.

Springer awards **"BestMasters"** to the best master's theses which have been completed at renowned Universities in Germany, Austria, and Switzerland. The studies received highest marks and were recommended for publication by supervisors. They address current issues from various fields of research in natural sciences, psychology, technology, and economics. The series addresses practitioners as well as scientists and, in particular, offers guidance for early stage researchers.

Yannik Venohr

Deep Learning in Personalized Music Emotion Recognition

Yannik Venohr
Applied and Computational Mathematics
Bergische Universität Wuppertal
Cologne, Germany

ISSN 2625-3577   ISSN 2625-3615   (electronic)
BestMasters
ISBN 978-3-658-46996-2   ISBN 978-3-658-46997-9   (eBook)
https://doi.org/10.1007/978-3-658-46997-9

This Springer Vieweg imprint is published by the registered company Springer Fachmedien Wiesbaden GmbH, part of Springer Nature.
The registered company address is: Abraham-Lincoln-Str. 46, 65189 Wiesbaden, Germany

If disposing of this product, please recycle the paper.

Acknowledgements

As I revisit this thesis that I wrote about a year ago, I find myself reflecting on the people who have supported me throughout this process. I would like to take this opportunity to express my gratitude to all those who have been part of this, directly or indirectly.

First and foremost, I would like to express my deepest gratitude to my supervisor, Prof. Dr. Matthias Ehrhardt, for his invaluable support throughout this project. I am especially grateful for your encouragement in allowing me to choose a topic outside your usual research area and for your belief in my potential at every step. Thanks to your guidance, this thesis has opened the door to the field of Music Information Retrieval, where I am excited to continue my academic journey as a PhD student. I am truly fortunate to have found a research area that combines my passion for music with my fascination for mathematics.

I also want to extend my gratitude to my friends, family and flatmates. Your support has kept me grounded during the inevitable ups and downs of writing a thesis. Whether it was providing a much-needed distraction, lending an ear to my frustrations, or simply sharing a laugh, you all played a crucial role in this process. Thank you!

Lastly, I want to thank the many great musicians whose work has been the soundtrack to my life and this thesis journey. Music has been my first love, a source of inspiration, and perhaps will be my last. I am profoundly grateful for the beauty it has brought to my life and the grounding presence it provided throughout this work.

Abstract

Music has the ability to evoke profound emotional responses in us. This master's thesis delves into the field of Musical Emotion Recognition (MER), intending to develop a mathematical model capable of predicting the emotional content of a musical piece. We explore the fundamentals of this interdisciplinary research area, including the relationship between music and emotions, mathematical representations of music, and the utilization of deep learning algorithms. Two MER models are developed and evaluated, one employing handcrafted audio features with a long short-term memory architecture and the other leveraging embeddings from the pre-trained music understanding model MERT released in March 2023. It is shown that, even with a basic regression head of fully connected layers, the MERT model can predict the emotional content of a musical piece. Moreover, the results indicate that utilizing MERT embeddings enhances predictions compared to traditional handcrafted features. Furthermore, driven by the inherent subjectivity of music, this thesis explores the notion of personalized emotion recognition. The findings suggest that personalized models surpass general MER systems' limitations and outperform even a perfect general MER system. This thesis concludes by opening up research opportunities to delve deeper into personalized MER.

Contents

Abbreviations

ADAM	ADAptive Moment estimation
AI	Artificial Intelligence
BI-LSTM	Bidirectional Long Short-Term Memory
CNN	Convolutional Neural Network
DEAM	Database for Emotional Analysis in Music
DFT	Discrete Fourier Transform
DL	Deep Learning
FC	Fully Connected
LLD	Low-Level Audio Descriptor
LSTM	Long Short-Term Memory
MAE	Mean Absolute Error
MER	Musical Emotion Recognition
MERT	acoustic **Music** und**ER**standing model with large-scale self-supervised **T**raining
MEVD	Music Emotion Variation Detection
MFCC	Mel Frequency Cepstral Coefficients
MIR	Musical Information Retrieval
ML	Machine Learning
MSE	Mean Square Error
ReLU	Rectified Linear Unit
RMSE	Root Mean Square Error
RNN	Recurrent Neural Network
SAM	Self-Assessment Manikin
SSL	Self-Supervised Learning
STFT	Short Time Fourier Transform
SVR	Support Vector Regression
VGG	Visual Geometry Group

Introduction

Music has the power to evoke strong emotions in us. It can bring us to tears, drive us into ecstasy, produce goosebumps or trigger vivid memories. It is often described as the universal language of emotions, capable of eliciting a wide range of feelings and moods in listeners. When people search for music on streaming platforms, they most frequently rely on emotional tags, highlighting the importance of emotion in music consumption [53]. Studies even suggest that the emotional effects of music are the most important reason why people engage in musical activities in the first place [13, 18]. Therefore, understanding and harnessing the emotional aspects of music are essential for enhancing music consumption experiences.

Research on the relationship between music and emotions dates back to the 1930s. Just about 20 years ago, the research field of Musical Emotion Recognition (MER) emerged as researchers began to study how to automatically recognize a musical piece's emotional content. MER is an interdisciplinary research problem that affects fields such as signal processing, machine learning, psychology and music theory.

In today's digital age, the importance of music in our lives applies even more. The contexts in which we interact with music are highly ubiquitous. With the advent of digital music streaming services, we can access extensive music collections at our fingertips. Conventionally, the management of music collections relies on catalog metadata, such as artist name, album name, and song title or is based on manual tagging of genres and emotions. As the amount of content continues to explode, these approaches are no longer sufficient. MER serves as one important building block of technologies that enable users to retrieve, organize and expand their music collections in a more **content-centric** manner.

A content-centric approach to music recommendation has the potential to promote fairness in the industry by suggesting music based on its ability to resonate with a listener's emotional state and taste rather than favouring already established

Y. Venohr, *Deep Learning in Personalized Music Emotion Recognition*, BestMasters, https://doi.org/10.1007/978-3-658-46997-9_1

artists based on metadata. Content-centric approaches can effectively bridge the gap between an obscure musician creating music in a Danish shed and a music enthusiast living in Chile, connecting them through their shared musical preferences.

In his extensive analysis of the relations between mathematics and music, David Benson concludes that mathematics has little to say about the "power of music as a medium of expression for moods and emotions" [5, p. xii]. However, researchers in the field of MER would beg to differ. MER utilizes mathematical methods to develop algorithms that predict the emotions a listener perceives in a musical piece. Traditionally, emotionally relevant features are extracted from audio based on domain knowledge, and machine learning algorithms like support vector machines are used to map these features to emotions. With the rapid development of artificial intelligence in the last years, a general shift toward using deep learning algorithms has been observed [22]. Architectures capable of capturing temporal dependencies, such as recurrent neural networks, have led to significant improvements. Even more recently, inspired by their success in natural language processing, the usage of large language models has recently extended to the field of MER. Researchers have utilized self-supervised pre-training on large unlabeled music corpora to build music understanding models that learn to create meaningful vector representations (embeddings) of a musical piece. In March 2023, for the first time, an acoustic **M**usic und**ER**standing model with large-scale self-supervised **T**raining (MERT) [34] was made publicly available to the research community, opening up many new research opportunities.

Most MER research focuses on creating general MER systems that predict the average emotion of a musical piece. This contrasts the fact that music perception is intrinsically subjective and influenced by many factors such as cultural background, age, gender, personality and training. For example, a person that listened to classical music in their childhood is more likely to perceive Bach as calming, while a person listening to Bach for the first time might perceive it as more activating. This subjectivity can be observed when comparing individual emotional responses to the same musical piece and is referred to as a low inter-rater agreement [18]. About a decade ago, Yang [53] proposed to address this issue through personalized models. Although some progress has been made in recent years, a significant research gap still exists [17].

Research Questions

Based on the previously described state of MER, this thesis aims to address the three following research questions (RQ):

- **RQ 1:** How can we develop a mathematical model that can predict the emotional content of a musical piece?
- **RQ 2:** Can utilizing embeddings from the pre-trained music understanding model MERT enhance the prediction results?
- **RQ 3:** Is it possible for this model to learn individual nuances when trained on personalized annotations? If so, how can we minimize the user burden of collecting individual annotations?

Outline

To address these research questions, this thesis is structured into eight chapters. The overall structure consists of two main parts: Foundations (Chaps. 2–5) and Development, Experiments, and Results (Chaps. 6–7).

Given the interdisciplinary nature of music emotion recognition, this thesis approaches the foundations from three perspectives. Chapter 2 introduces the field of MER, by exploring various aspects of the relationship between emotions and music. Chapter 3 bridges the gap between the *sound of music* and a mathematical representation of it that can be understood by computers. Chapter 4 describes the methods employed to enable machines to learn. These foundational concepts are then consolidated in Chap. 5, which provides an overview of current research in MER.

Building upon this groundwork, two distinct architectures for MER models will be designed and implemented. Chapter 6 presents the details of model development, including design choices and the learning pipeline. In Chap. 7, we will assess these models' capabilities and address subjectivity by conducting experiments on the personalization of MER-systems.

Finally, Chap. 8 reflects on this introduction, summarizing the answers to the research questions and providing an outlook on future research opportunities.

Music Emotion Recognition

The relationship between music and emotion has captivated scholars across various disciplines, such as philosophy, musicology, psychology, biology, anthropology, and sociology. Regardless of the academic field, "the notion that music expresses emotion has a venerable history, and its validity is rarely debated" [53, p. 4].

As Stevie Wonder sings in *Sir Duke*, music can be described as "a language we all understand". As a matter of fact, several cross-cultural studies indicate that there are music psychological and emotional cues in music that can transcend the limits of language and culture [56]. Western listeners, for example, can detect the expressed emotions in Hindustani ragas "even though they are unfamiliar with the tonal system and the raga-rasa system of conveying moods within that tonal system" [4, p. 57].

This evidence lends support to the idea that machines, too, have the potential to pick up on these cues and capture the nuanced emotions conveyed in music. The research field that explores computational models for detecting emotions in music is called **Musical Emotion Recognition (MER)**. This chapter aims to briefly introduce all the relevant aspects of MER for this thesis.

Section 2.1 will present the necessary preliminary knowledge for MER, establishing the foundation for understanding the modelling of emotions in music within this thesis. It will define the concept of emotions in music, highlighting the key features and characteristics that will be considered during the modeling process.

Section 2.2 will critically examine the challenges associated with defining a ground truth for a MER model, taking into account the inherent subjectivity involved in interpreting and perceiving emotions in music and presents methods of personalization to address this issue.

Section 2.3 introduces the Database for Emotional Analysis in Music (DEAM), which will serve as the ground truth in this thesis. It will outline the structure and properties of the dataset.

Y. Venohr, *Deep Learning in Personalized Music Emotion Recognition*, BestMasters, https://doi.org/10.1007/978-3-658-46997-9_2

2.1   Domain Definition

In accordance with the MER research framework proposed by Han et al., [22] we will start with a domain definition. In the following, we will present the necessary preliminary knowledge and define the conceptual framework this thesis will use to model emotions in music.

Expressed, Perceived and Induced Emotions
An important distinction in analyzing emotions in music is that between expressed, perceived and induced emotions:

- The **expressed emotion** refers to the emotion the performer intends to communicate with the listeners by choosing certain musical elements such as melody, harmony, rhythm, lyrics or dynamics.
- The **perceived emotion** is the emotion the listener perceives while listening to the music or interprets through the musical properties of a piece.
- The **induced emotion** refers to the emotion the listener actually feels or shows as psycho-physiological responses as a result of listening to the music. While a song may be perceived as happy due to its tempo, lyrics and harmonies, it can still induce sadness by triggering a specific memory in the listener. An example is the song *Happy Birthday*. While most people would agree to perceive the melody as rather happy (also being written in a major key, which in Western music is usually perceived as happy), for a specific person, this song might induce sadness, triggering the memory of a lost partner.

While the expressed emotion can only be defined by the musician himself or herself, the latter two refer to the listeners' emotional responses and thus are more relevant for this thesis. Both perceived and induced emotions are highly dependent on the listener's musical, personal, and situational factors and, therefore, subjective (see Sect. 2.2). Still, most MER research and this thesis describe the perceived emotion due to the fact that it is less dependent on individual experiences and context compared to the induced emotion [53].

Emotion Models
Many scholars in the field of music psychology have studied emotions in music and tried to find models to describe these emotions. As a result, two dominant views or taxonomies for emotion representation were developed:

- **Dimensional models** are models where emotions are identified based on their location in an emotional space with a small number of emotional dimensions. In 1980 Russel [45] introduced the circumplex model of affect, a two-dimensional emotion model. One is the valence dimension ranging from unpleasant to pleasant, and the other is the arousal dimension ranging from deactivation to activation. Figure 2.1 shows a visual representation of this emotional space. Some models use Russel's two dimensions and add a dominance dimension. Due to experiments showing a high correlation between the dominance and arousal dimension and the higher complexity in the annotation process, this dimension is rarely used in research [21].

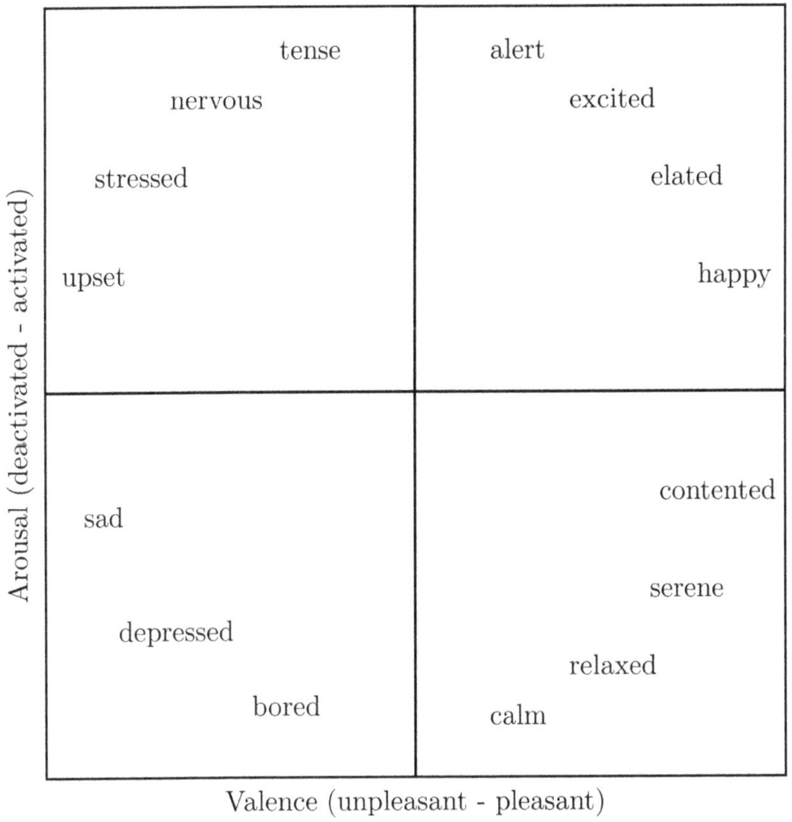

Fig. 2.1 Russels circumplex model (labels as in [44])

Another variant of Russell's model is Thayer's model. The author suggests that two basic dimensions of describing emotions are two separate arousal dimensions: energetic arousal and tense arousal [21]. In this model, valence can be described as a combination of energetic and tense arousal. In current MER research Russel's model is the most commonly used emotion model [22].

- **Categorical models** are models where emotions are described with a discrete number of classes of adjectives. Hevner's affective cycle [25] is one of the earliest and most influential categorical music emotion models. In 1935 Hevner found 67 emotional adjectives to describe the emotional space expressed by music. Furthermore, she divided these 67 emotional adjectives into eight categories: dignified, sad, dreamy, serene, graceful, happy, exciting, and vigorous. Another approach comes from research on facial expressions. Ekman [14] found six basic emotions (anger, fear, sadness, enjoyment, disgust, and surprise) and showed that these emotions are cross-cultural. Another popular emotion set used in MER categorizes emotions in music in classes: happy, angry, sad, and relaxed. The upside of these four categories is that they correspond to the four quadrants of Russel's dimensional model.

Compared to dimensional models, there are some advantages and disadvantages of using categorical models. On the one hand, categorical models can have the capability of describing more complex emotions, such as *longing* or *gracefulness*. On the other hand, two main issues come with categorical models. The issue of **ambiguity** and the issue of **granularity**.

Especially models with many categories have shown to be problematic in annotation due to their high ambiguity [42]. Russel states that a "human being usually is able to recognize emotional state but has difficulties with its proper defining" [53, p. 6]. Even when there is an agreement between listeners, there is often ambiguity in the terms used for describing a certain emotion. Or, the other way around, people associate different emotions with the same terms. One obvious approach to reducing ambiguity is to reduce the number of categories.

Reducing the number of categories leads to the issue of granularity. Categorical models with a low number of categories are limited in their resolution by their number of categories. Thus they are not capable of describing more subtle shades of emotions. "This is undesirable because a retrieval system with limited emotion vocabulary may not satisfy the user requirement in real-world music retrieval applications" [53, p. 6].

When using emotion models based on a dimensional approach, the granularity and ambiguity issues no longer exist because no ambiguous classes are needed, and the emotion plane implicitly offers infinite emotion descriptions. Thus this model has been proven simple, highly reliable and easily understood by experiment participants and annotators [53, p. 6]. Therefore this thesis will also use the valence and arousal plane proposed by Russel [45].

Static versus Dynamic
MER research and systems can additionally be divided into two categories:

- **Static** or song level MER tries to assign an overall emotion label to an entire musical piece. This is relevant for using MER in music recommendation.
- **Dynamic** MER, also referred to as Music Emotion Variation Detection (MEVD) focuses on the time-dependent and dynamic nature of music. In this approach, emotion labels are considered as changing throughout a composition. Common intervals for this are 0.5 to 1 second. Commonly MEVD is combined with a dimensional emotion model. Due to a lack of datasets, research on dynamic MER using a categorical model is almost non-existent [22].

In a survey, Han et al. [22] observe a shift in research towards dynamic MER. According to him, this can be explained by two reasons. Firstly, dynamic processing is more in line with the characteristics of music. Some researchers argue that "due to the complexity and temporal variation of emotions in music, it may be ambiguous and inaccurate to mark a piece of music with one single annotation" [59, p. 1]. Secondly, with the emergence of sequence models like recurrent neural networks, dynamically recognizing continuous emotions has become more convenient.

However, in music recommendation, users are typically interested in receiving recommendations for complete songs rather than music segments. It is also less complex than dynamic MER, making it more practical for implementation. Thus this thesis will focus on static MER.

Summary: Domain Definition
This thesis will use Russel's **dimensional** emotion model to do **static** MER of the **perceived** emotion of a musical piece.

2.2   Subjectivity and Personalization

To train any automatic model to recognize emotions in music, there needs to be some definition of the *actual* perceived emotion of a music piece. This is usually called the **ground truth** to be modelled. Typically this is done by inviting subjects to annotate the emotion of a music piece, usually referred to as **annotators**. This section discusses some issues associated with this process and discusses ways to address these issues.

Issues Defining Ground Truth
In order to be able to define the ground truth, the annotator would have to be able to tell what the *actual* perceived emotion of a music piece is.

This is impossible because music perception is intrinsically **subjective** and influenced by many factors such as cultural background, age, gender, personality and training. For example, the author of this thesis has mostly listened to Western music in his life. Thus he would perceive the emotion of Indian music differently compared to a person that has listened to Indian music their entire life. Or a person that listened to classical music in their childhood is more likely to perceive Bach as calming, while a person listening to Bach for the first time might perceive it as more activating.

This can be observed when examining annotations from different subjects on the same song. Figure 2.2 shows valence and arousal annotations by different subjects for two songs from the Database for Emotional Analysis in Music (DEAM) (see Sect. 2.3). While it is possible to see the tendency, that Song 2 is perceived as less activating and slightly sadder than Song 17. It is still apparent that there is a large statistical dispersion between the annotations for the same song for both valence and arousal. This is also referred to as a **low inter-rater agreement** and will be further discussed in Sect. 2.3.

An additional issue in the annotation process is the heavy cognitive load it imposes on the subjects. It has been found that rating emotion in a continuum, using either an ordinal rating scale or a graphic rating scale, both impose a high cognitive load on the annotators. Thus it is even harder to ensure consistent rating scales between different subjects and within the same subject [53].

Even though, as stated earlier, there are cross-cultural emotional cues in music; still, a big part of the emotional response is triggered by culture-specific cues [4]. This introduces another issue to be considered when developing a MER system: the issue of **bias**. Most datasets show a strong bias in the selection of the annotators. For example, most annotations are carried out by WEIRD participants and researchers (from Western, Educated, Industrialized, Rich, and Democratic backgrounds) [17].

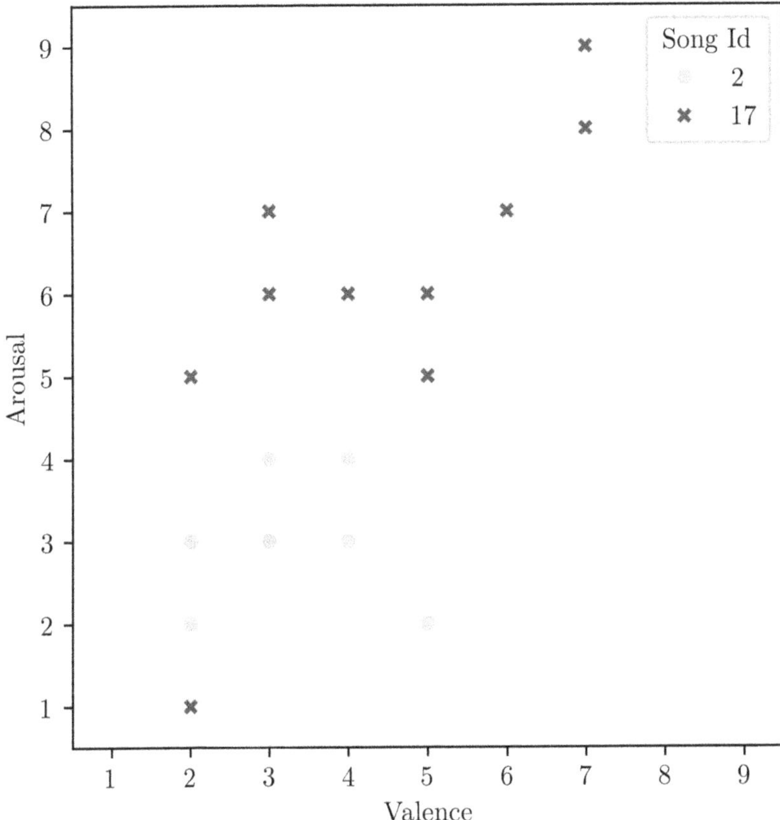

Fig. 2.2 Distribution of valence and arousal values for song 2 (*Tonight A Lonely Century* by *The New Mystikal Troubadours*) and 17 (*Gtr Fills Mix* by *Phoephus*) in the DEAM dataset.

As a consequence, certain cultural-specific aspects of emotions in music are not well represented. This should always be considered in the selection process of the annotators.

Strategies to Address Subjectivity
As mentioned before, emotions in music are subjective, and even the same subject may provide different annotations for the same musical piece in different contexts.

This subjectivity and variability pose a significant challenge for building accurate MER systems.

Ultimately, while we may never achieve a perfect ground truth, we can strive to define a *sufficiently enough* ground truth that balances accuracy and feasibility. In this chapter, we will explore some strategies that can improve the quality of the ground truth and their implications for the design and evaluation of MER systems.

Typically the emotion of a music piece is annotated by more than one subject. Psychology literature suggests that a song has to be annotated by at least 30 subjects for the annotation to be reliable [53]. This is rarely the case in practice due to the high administrative effort involved in this process. When several annotators are involved, the ground truth is typically defined by the mean or the median of all ratings. This can be problematic because "it unintentionally creates the mirage of potential universality" [17, p. 7] even though many datasets have been reported to have a low statistical inter-rater agreement. As a result, even a perfect MER system is merely correctly predicting the average annotation and not the perception of an individual user. Systems using this approach of averaging annotations will be referred to as **general** MER systems.

Another common approach is having musical *experts* annotate to gain more consensus on the results. This practice is seen as problematic for some use cases because the contexts in which musical experts experience music are quite different from those of non-experts [53]. It is argued that since the desired MER system is expected to be used in the everyday context, the annotation should also be carried out by *ordinary* people.

One approach to limit the cognitive load on the subjects is to reduce the length of the music pieces. This has several effects. On the one hand, the expression of a shorter segment tends to be more homogeneous, resulting in higher consistency in the individual listener's ratings. On the other hand, if the segment is too short, the listener cannot hear enough of it to accurately determine its emotional content [53]. Empirical studies on which length of music best presents a stable mood state for classical music found that 6-8 seconds serve as the most stable value. Again, when trying to determine the mood of an entire song, the segment needs to be representative of the entire song. A segment length commonly used in literature is 20–30 seconds. This presents a trade-off between having a long enough segment to capture the song's mood and reducing the listener's cognitive load.

Personalization

An approach to avoid the shortcomings of general MER systems and to embrace subjectivity, rather than seeing it as a problem, is personalization. There are two common approaches to doing this:

- In **personalized** MER a model is trained using the target user's independent annotations. This alleviates many of the above-stated problems and requires many annotations from the listener for the system to work.
- In **groupwise** MER groups of similar users with characteristics such as demographics, academic background, language, music experience or personality are formed and averaged. The idea of groupwise models is to be able to make better recommendations just based on user metadata.

It has been observed that personalized models significantly improve performance over general models, while groupwise models do not [17]. However, due to the high cognitive load, it is difficult to gather many personal annotations within a short time and thus making this approach difficult for many real-world applications. Just imagine having to annotate 500 songs for a music recommendation system to work correctly.

Therefore research has focused on combining the strength of general MER systems (data availability) and the strength of personalized MER systems (taking subjectivity into account).

The most intuitive approach to do this, introduced by Yang [53], is the **model averaging approach**. This approach combines predictions of a general model trained on annotations from a broad user base and a personal model trained on annotations by the target user. Usually, they are averaged with a weighting factor, which we will refer to as the **degree of personalization**.

The **residual modelling approach** also uses two models [54]. In addition to the general MER model, another model is trained to predict the residual between the emotion of the general model and the individual emotion. The final prediction of the residual modelling approach is the sum of these two models. With this, the authors could improve the performance of a general model with just ten individual annotations.

Gomez-Canon et al. concludes "while recent efforts have been made in this direction, more future work is needed" [17, p. 14].

2.3   Database for Emotional Analysis in Music

This chapter introduces the Database for Emotional Analysis in Music (DEAM) that will serve as the ground truth for this thesis.

Audio copyright restrictions are a big issue regarding datasets for MER research. Due to these restrictions, some researchers resort to creating their own datasets, which are not publicly available. Thus making different MER algorithms hard to compare. However, for researchers to develop better MER systems, it is necessary to compare algorithms and replicate research results.

The biggest effort to develop a benchmark for MER research has taken place in the *Emotion in Music* task organized by MediaEval.

> "MediaEval is a benchmarking initiative dedicated to evaluating new algorithms for multimedia access and retrieval. [...] MediaEval attracts participants who are interested in multimodal approaches to multimedia involving, e.g., speech recognition, multimedia content analysis, music and audio analysis, user-contributed information (tags, tweets), viewer affective response, social networks, temporal and geo-coordinates" [MediaEval].

MediaEval organized the *Emotion in Music* task from 2013 to 2015. Each year they provided a development set (audio + annotations) and a test set (audio). Research teams were then invited to submit their predictions for the test set. The benchmark attracted, in total, 21 research teams to participate in the challenges.

DEAM is the combination of the data sets developed in these three years. Aljananaki et al. [3] describe the benchmark and dataset's design and analyzes the research teams' approaches.

Music Database

The database consists of royalty-free music from several sources, including freemusicarchive.org (FMA), jamendo.com and the medleyDB dataset. There are 1,744 excerpts of 45 seconds. This makes DEAM the biggest database for dimensional MER [22]. There is a wide variety of genres, including rock, pop, soul, blues, electronic, classical, hip-hop, international, experimental, folk, jazz, country, world, rap, reggae and pop. The distribution of genres for the 2014 development set is presented in Table 2.1. The audio files have been manually checked to exclude recordings with bad audio quality or files containing speech or noise. For each artist, they selected no more than five songs to be included in the dataset.

Annotations

It is essential to mention that DEAM was mainly developed as a benchmark for dynamic MER, but in addition to dynamic annotations, the database also contains static annotations. These will be used in this thesis.

The crowdsourcing platform Amazon Mechanical Turk has been used to collect annotations. The annotators passed a test with multiple choice and free-form questions to demonstrate a thorough understanding of the task and the ability to produce good quality work. Due to the fact that royalty-free music is usually not well known,

Table 2.1 Distribution of genres for 744 songs of the 2014 development set

Genre	Count
Classical	116
Country	105
Jazz	105
Blues	100
Electronic	92
Pop	78
Rock	77
Folk	71

the music was mainly unfamiliar to participants. Only one per cent of participants reported knowing a song.

The dynamic annotations were collected with a slider in a web interface on a scale from −10 to 10 while the song was being played. Valence and arousal were collected separately to reduce the cognitive load for the subjects. After the dynamic annotations, the static annotations were made on a nine-point scale on valence and arousal for the whole excerpt. For this, the Self-Assessment Manikin (SAM) as proposed by Bradley and Lang [6] has been used. SAM is an efficient cross-cultural measurement of emotional response, where Russel's emotional dimensions are represented as graphic characters along a scale. For valence SAM ranges from a frowning, unhappy character to a smiling, happy character. For arousal, SAM ranges from sleepy with eyes closed to excited with eyes open (see Fig. 2.3).

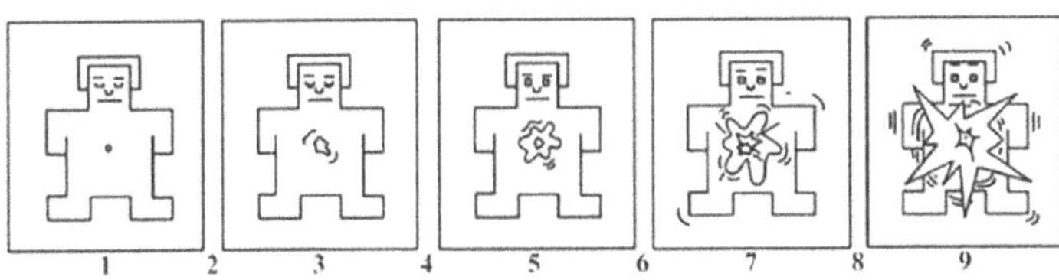

Fig. 2.3 Self-Assessment Manikin (SAM) for the arousal dimension as in [48] (SAM [6])

While at least ten subjects have labeled each song, there were no restrictions on the number of songs each subject labeled. This lead to the fact, that there are some subjects, that labeled many songs, while others just labeled a few. This is visualized in in Fig. 2.4. 34 subjects have annotated at least 100 songs, and 10 have annotated at least 500 Songs. This makes this dataset interesting for personalized MER.

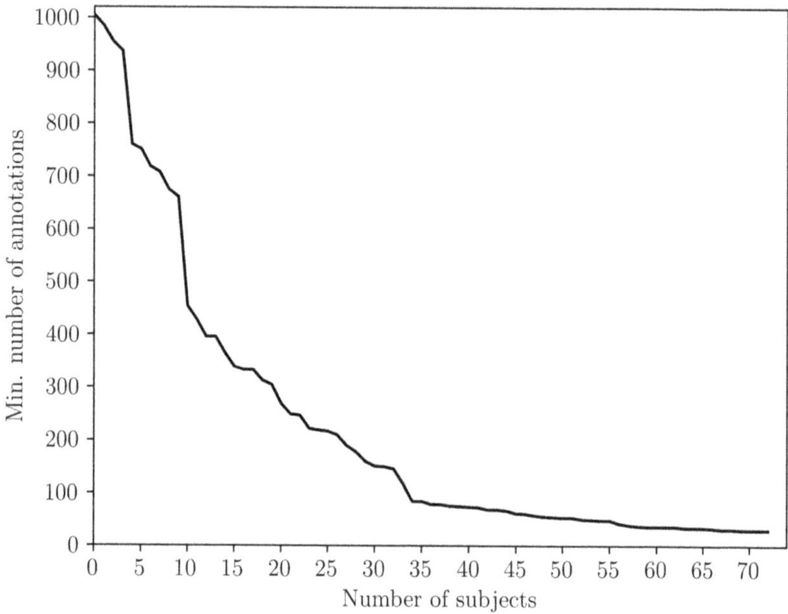

Fig. 2.4 Distribution of the number of annotations per subject

Data Structure

In the database, we have the following files and folders that are relevant to this thesis:

- **Audio files:** A folder with all separate audio files stored as `[songID].mp3`.
- **Audio features:** A folder with separate files stored as `[songID].csv` containing audio features for this song (see Sect. 3.3).
- **Annotations averaged per song:** A table with columns for songID, valence mean, valence standard deviation, arousal mean and arousal standard deviation. This table has 1744 entries.
- **Individual annotations:** A table with columns for workerID, songID, valence and arousal. This table has 17464 entries.

Data Distribution

Let us look at the data distribution for the annotations in DEAM. Firstly we will visualize the individual annotations. Figure 2.5 shows the dataset's distribution of valence and arousal values. It can be observed that the majority of the data points tend to cluster around the middle region of the plot, indicating a prevalence of medium valence and medium arousal responses. The average valence and arousal of all 17646 annotations are 4.9 and 4.8.

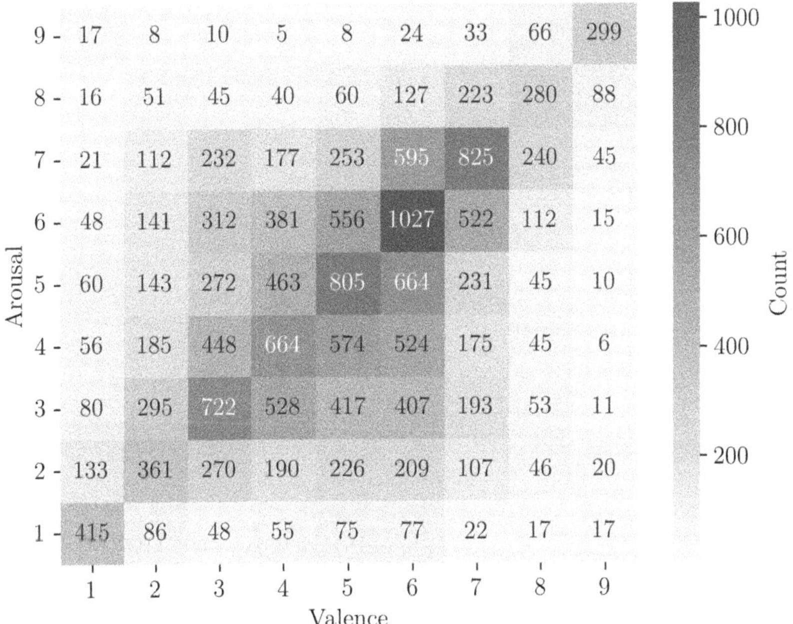

Fig. 2.5 Distribution of valence and arousal per rater

Additionally, it is worth noting that a substantial number of annotations align along the diagonal of the plot. This observation raises a question regarding a potential correlation between valence and arousal. The presence of annotations along the diagonal could suggest that as valence increases, so does the level of arousal, and vice versa. Another explanation could be that songs with a perceived emotion in the upper right and the lower left quadrant have been overly represented in the dataset.

For the *Emotion in Music* task, the ground truth was defined as the mean of all annotations for one song. The distribution of valence and arousal averaged per song can be seen in Fig. 2.6. The averaging procedure has contributed to a stronger centralization of the annotations. Calculating the average values mitigates the extreme variations and outliers within individual annotations, leading to a representation that tends to cluster around the centre of the emotional model.

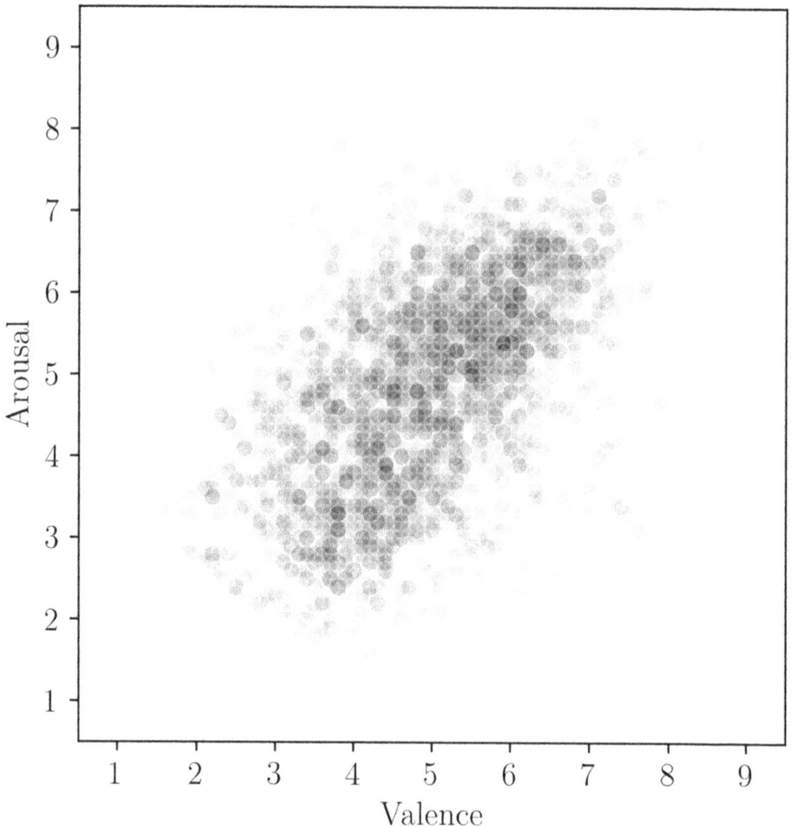

Fig. 2.6 Distribution of valence and arousal averaged per song

Finally, we will inspect the **inter-rater agreement**. It describes the degree of agreement among single annotators about the emotion of a song. As Fig. 2.2

exemplified for two songs, the annotations for the same song can differ quite strongly between different subjects. To analyze this more systematically, we can have a look at the standard deviation σ_j of valence or arousal ratings for a song j. We define it as

$$\sigma_j = \sqrt{\frac{1}{N}\sum_{i=1}^{N}(x_{i,j} - \overline{x}_j)^2}, \text{ where } \overline{x}_j = \frac{1}{N}\sum_{i=1}^{N}x_{i,j},$$

N is the number of annotations for song j and $x_{i,j}$ is the annotation for song j by subject i for either valence or arousal.

Figure 2.7 shows the distribution of σ_j for $j = 1, 2, \ldots, 1744$ for valence and arousal. In general, one can notice that the inter-rater agreement for arousal is slightly higher than for valence. The highest standard deviation is 2.9 for valence and 2.59 for arousal. Considering a rating scale from 1 to 9, this is quite high. For 75% of the songs, the ratings for valence and arousal have a standard deviation of more than ≈ 1.25. When normalizing to a scale of 0 to 1, this is equivalent to 0.14. This will become relevant when discussing the results of the general MER model in Sect. 7.2.

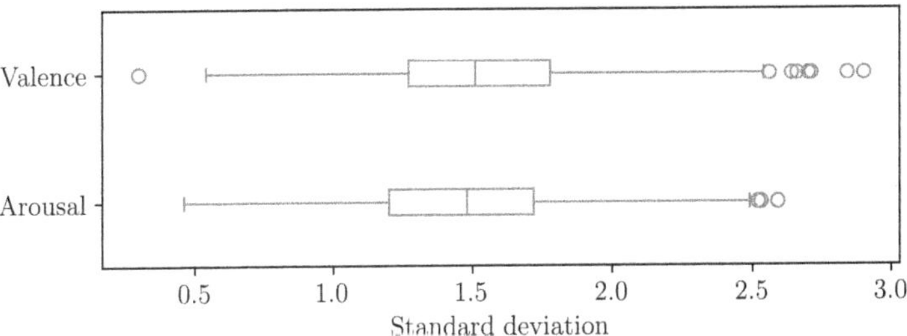

Fig. 2.7 Standard deviation of ratings between subjects for valence and arousal

Describing Music Mathematically 3

For humans, music and sound can be experienced and appreciated on an emotional and subjective level. In the previous Chapter we introduced a way of representing emotions mathematically. In this Chapter, we will learn about different mathematical representations of sound an music.

Section 3.1 describes the necessary foundations for a digital representation of sound and introduces the time-pressure representation of it.

Section 3.2 introduces the Fourier transform that enables us to transform the time-pressure into a time-frequency representation of sound.

Section 3.3 describes methods to extract other emotionally relevant features of an audio signal and gives an overview over the feature set, that is provided in the Database for Emotional Analysis in Music (DEAM).

3.1 Sound to Digital Music

Firstly we need to understand the steps of getting from *the sound of a person playing the guitar* to a vector representation of this information.

Describing music or sound mathematically involves understanding the physical properties of sound waves and how they interact with the environment. Sound waves are created when an object, such as a guitar string, is set into motion and begins to vibrate. This vibration causes the air molecules around the object also to vibrate, causing displacements and oscillations of air molecules, resulting in local regions of increased and decreased pressure [39]. These disturbances of pressure travel through the air as a longitudinal wave.

The pressure wave can be described by plotting the changes in pressure at a particular position (e.g. the microphone) over time. This is known as a **pressure-time** plot, also referred to as the **waveform** of the sound. The pressure-time plot

© The Author(s), under exclusive license to Springer Fachmedien Wiesbaden GmbH, part of Springer Nature 2025
Y. Venohr, *Deep Learning in Personalized Music Emotion Recognition*, BestMasters,
https://doi.org/10.1007/978-3-658-46997-9_3

shows the pressure wave amplitude on the y-axis and time on the x-axis. Figure 3.1a shows the waveform of the brass section intro of Stevie Wonder's *Sir Duke*. Between seconds 8 and 9, for example, we can see the part where the drums are playing four even beats on the bass drum, and the brass section does not play. When kicking the base drum, the air is compressed. This compression can be seen as deflections in the waveform. We can see four spikes in the amplitude for the four kicks, with a lower amplitude in the pauses in between. We perceive the amplitude as the **loudness** of as sound.

(a) First 10 seconds: the brass section intro

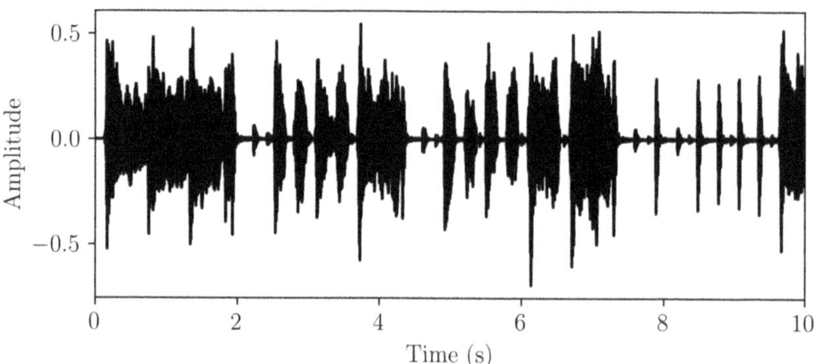

(b) The beginning of the first note of the intro

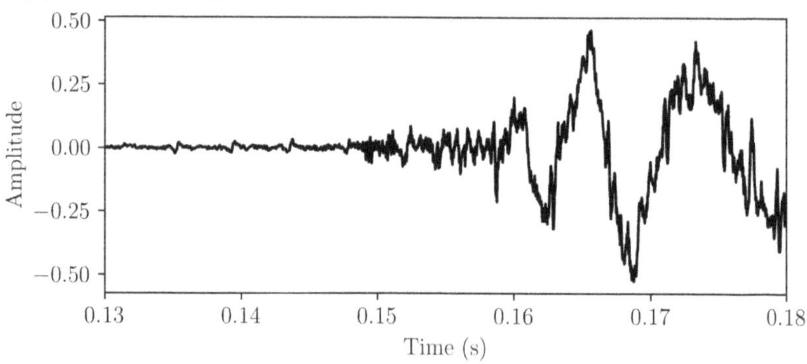

Fig. 3.1 Waveform of Stevie Wonder's *Sir Duke* on different time scales

If we take a closer look at the beginning of the first note of the brass section (Fig. 3.1b), we can discover some other attributes of sound. We can observe that increased and decreased air pressure points have some form of repeating patterns (periodic). The prototype and the simplest form of a periodic wave is the sine wave, which can be completely described by its amplitude, its frequency and its phase (starting point). The frequency in which these patterns repeat themselves is perceived by us as the **pitch** of a sound and is measured in **Hertz** (Hz). The audible frequency range for humans is between about 20 Hz and 20 000 Hz [39]. Two frequencies are perceived as similar if they differ by a power of two. This is what motivated the concept of an octave. For example the note A3 has a frequency of 220 Hz, while the note A4 (one octave higher) has a frequency of 440 Hz. In other words, "the human perception of pitch is logarithmic in nature" [39, p. 22].

In Fig. 3.1b, we can also see that, in practice, the waveform does not look like a pure sine wave. Instead, it consists of many different frequencies at the same time. When the brass section is playing, what we perceive as a B, it creates air displacements in many different frequencies at the same time. The frequency of the B (246.94 Hz), which we perceive as the frequency that defines the note, is called the **fundamental frequency**. The way in which other frequencies are present and the way the composition of frequencies changes over time is called **timbre** or **tone colour** of the sound. Timbre is the sound attribute that allows listeners to distinguish that the B is being played by the brass section instead of a piano.

The analog audio signal can be described as a function $f : \mathbb{R} \mapsto \mathbb{R}$ mapping from time to amplitude. When storing and working with sound digitally, we are limited to a finite number of parameters that can be stored and processed. Thus we need to **digitize** the audio signal by using an **equidistant sampling**. We do this by defining a function $x : \mathbb{Z} \mapsto \mathbb{R}$ as

$$x(n) = f(nT)$$

where $T \in \mathbb{R}^+$ is called the **sampling period** and $x(n)$ is called the **sample** taken at time $t = nT$. The inverse

$$F_s = \frac{1}{T}$$

of the sampling period is also called the **sampling rate**. The **Nyquist theorem** states that the original analog signal f can be reconstructed perfectly from its sampled version x if f does not contain any frequencies higher than half of the sampling rate F_s [5]. This explains the sampling rate of 44 100 Hz used for CDs, which is slightly more than double what the human ear can hear.

Using Torchaudio [57], a library for audio and signal processing with PyTorch, we can examine the digital version of *Sir Duke*.

```
import torchaudio
SAMPLE_WAV = ".../data/Stevie-Wonder-Sir-Duke.wav"
metadata = torchaudio.info(SAMPLE_WAV)
waveform, sample_rate = torchaudio.load(SAMPLE_WAV)
```

Code example 3.1 Using Torchaudio to examine a .wav file

First, let us look at its `metadata`. In this case, the song has a sampling rate of 48 000 Hz. Since it has a length of 3min 52s, it consists of 11 160 000 frames ($232 \cdot 48000 \approx 11160000$). The song is recorded in stereo. Thus, we have two channels, in which each frame consists of 16 bit encoded as signed integer linear Pulse-code modulation. So `waveform` is a tensor of shape (2, 11160000). For our purposes, it is sufficient to reduce the stereo recording to mono [53]. This can be achieved by calculating the mean of the left and right channels.

To conclude, when we talk about the raw audio of a song with length l in seconds in this thesis, we refer to a vector of $\mathbb{R}^n$, where $n = F_s \cdot l$. In this case, the entry at index i of our vector is a value correlating with air pressure at time $t = i \cdot T$ in seconds. To be precise, we do not actually have values of entire $\mathbb{R}$ since we are working with encoded digital data. Rather we have a value range limited by the 16bit signed integer linear encoding. For practical reasons, we will note it as values of $\mathbb{R}$.

3.2   Time-Frequency Representation

In the Sect. 3.1, our mathematical representation of sound was linked to the raw audio signal, and we were thinking about it as represented in a pressure-time plot. Also, we have seen that the raw audio signals comprise many different periodical signals. This Section introduces another way of representing audio signals, called the **time-frequency representation**. This representation is closer to how we perceive music and the basis for most digital music processing, which will be presented in Sect. 3.3. We will outline the mathematical ideas for obtaining this representation in the following. The ideas presented are based on the reference work *Fundamentals of Music Processing* by Meinard Müller [39].

Fourier Analysis

As stated in the previous Section, the prototype of a periodical signal is the sine function (sinusoid). For now, let us imagine sinusoid waves of a particular frequency, amplitude and shift as building blocks. Every signal can be built by stacking these building blocks on top of each other. In mathematical terms, this is known as the Fourier series for non-periodic functions defined on a fixed interval. It is usually more convenient to work with these building blocks for analyzing and processing tasks than with the raw audio signal. The procedure to obtain these building blocks is the **Fourier analysis**.

The main idea of Fourier analysis is to compare the signal with sinusoids of various frequencies $\omega \in \mathbb{R}$ (measured in Hz). Each sinusoid is one of our previously mentioned building blocks. After comparing the similarity of our signal to each of these sinusoids, we get an amplitude coefficient $d_\omega \in \mathbb{R}_{>=0}$ for each $\omega \in \mathbb{R}$. A large d_ω means a high similarity of our signal and sinusoid. Doing so can break up a signal into its frequency components.

So firstly, we would like to know the similarity between our signal f and a sinusoid $g = \cos_{\omega,\varphi}$ of a given frequency ω and phase φ (measured in normalized radians). As mentioned in the previous Section, an analog signal can be described as a function $f : \mathbb{R} \mapsto \mathbb{R}$. One way to measure the similarity between two functions is to ensure both functions have the same sign over time. If f assumes positive values, so should g, and vice versa. When multiplying values of the same sign, we get a positive value; if we multiply values of a different sign, we obtain a negative value. Due to that, we can rephrase the above by saying that f and g are similar if their product is positive at *as many time steps as possible t*. To make *as many time steps as possible* more defined, we will consider infinitesimal small time steps and use the integral:

$$\int_{t\in\mathbb{R}} f(t)\cos_{\omega,\varphi}(t)dt.$$

Now that we have a way to measure how similar our signal is to $\cos_{\omega,\varphi}$, we can define d_ω and φ_ω for each frequency ω as

$$d_\omega := \max_{\varphi\in[0,1)}\left(\int_{t\in\mathbb{R}} f(t)\cos_{\omega,\varphi}(t)dt\right).$$

$$\varphi_\omega := \arg\max_{\varphi\in[0,1)}\left(\int_{t\in\mathbb{R}} f(t)\cos_{\omega,\varphi}(t)dt\right).$$

This form does feel a little bit awkward since it involves optimization. Using the concept of complex numbers, we can achieve a more elegant formulation of the Fourier transformation. The principal idea is to use our two coefficients d_ω and φ_ω as polar coordinates for a single complex number that we call $c_\omega \in \mathbb{C}$. By making use of several trigonometric identities, it can be noted as

$$c_\omega = \int_{t \in \mathbb{R}} f(t) \exp(-2\pi i \omega t) dt.$$

Now, we can stack our building blocks on top of each other and represent our signal as a combination. In mathematical terms, this means we superimpose the sinusoids of all possible frequency parameters $\omega \in \mathbb{R}$, each weighted by the respective coefficient d_ω and shifted by φ_ω. If we do this, we can note our original signal f as an integral, that we call the **Fourier representation** of the signal as

$$f(t) = \int_{\omega \in \mathbb{R}} c_\omega \exp(2\pi i \omega t) d\omega.$$

Discrete Fourier Transform

As discussed in Sect. 3.1, when working digitally, we are limited to a finite number of parameters that can be stored and processed. Thus integrating ω over $\mathbb{R}$ is not an option. However, since we are working with digitized, thus discrete signals, we can utilize the **Discrete Fourier Transform (DFT)**.

Analog to how we digitized our signal in Sect. 3.1, we can sample our frequency axis by just considering frequencies $\omega = \frac{k}{M}$ with $M \in \mathbb{N}$ and $k = 0, 1, ..., N - 1$. For practical reasons, the frequency resolution M is usually set equal to the number of samples in our digital signal N. We can note the discrete Fourier transform as

$$X(k) = \sum_{n=0}^{N-1} x(n) \exp\left(\frac{-2\pi i}{N}\right) d\omega$$

for integers $k = 0, 1, ..., M - 1 = 0, 1, ..., N - 1$. $X(k)$ is the Fourier coefficient for $\omega(k) = \frac{k}{N}$.

Two hundred years ago, Gauss and Fourier discovered an algorithm that "exploits redundancies across sinusoids of different frequencies to jointly compute all Fourier coefficients by a recursion" [39, p. 52], called the **Fast Fourier Transform**. This algorithm makes the otherwise computationally complex procedure a lot more efficient and changed whole industries and is now being used in billions of telecommunication and other devices.

Using this algorithm, we can calculate the DFT of the beginning of the first note of the brass section. One way to vizualize this is to plot the Fourier coefficient $X(k)$ over the frequency ω. This is shown in Fig. 3.2.

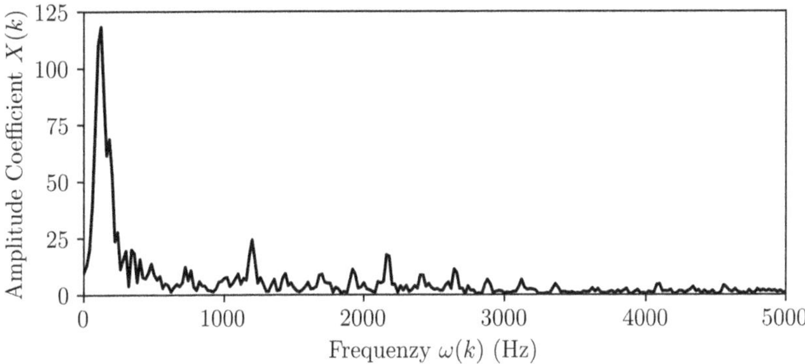

Fig. 3.2 Discrete Fourier Transform of the first note ($0.13s$—$0.18s$) of Stevie Wonder's *Sir Duke*

Short-Time Fourier Transform

Previously, we introduced an efficient way to calculate the Fourier Transform of a digital audio signal and thus gain insight into the frequencies a signal is made up of.

> "The signal [time-pressure representation] tells us when certain notes arc played in time but hides the frequency information. In contrast, the Fourier transform of music displays which notes (frequencies) are played but hides the information about when the notes are played" [39, p. 41].

The quote above shows the limitation of the Fourier transform for music pieces that consist of more than one note (so most music pieces). Since the Fourier transform yields frequency information that is averaged over the entire time domain, we lose the information of the time instance when a note is played.

We will now introduce the **Short Time Fourier Transform (STFT)** to address this issue. The main idea of the STFT is to consider only a tiny section, instead of considering the entire signal. The idea behind the STFT is to break down the signal into short overlapping windows and to compute the Fourier Transform for each window.

This results in a **time-frequency representation** of the signal, where we can see how the frequency content changes over time. This can be visualized as a (logarithmic) **spectogramm** (see Fig. 3.3). The x-axis is plotted on a logarithmic scale due to previously discussed logarithmic nature of sound. This visualization enables us to see the three ascending notes the brass section starts with. While we perceive the B as being the fundamental frequency of the first note, we can see that the tone actually consists of many different frequencies at the same time.

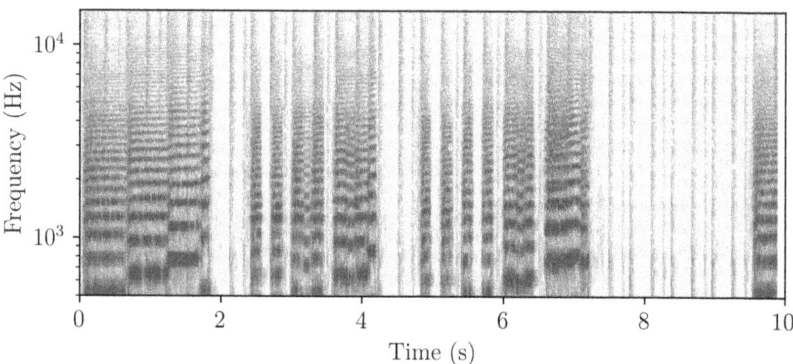

Fig. 3.3 Spectogramm of the intro of Stevie Wonder's *Sir Duke*

3.3   Audio Features

As in many other machine learning domains, it is common to prepare the data in a certain way before feeding it to the machine learning model. This process is called **feature extraction**.

Concerning feature extraction in this thesis, an important distinction has to be made:

- **Handcrafted** feature extraction involves the manual design and selection of features from raw data, requiring domain knowledge and expertise in the specific problem domain. This is the scope of this Section.
- Feature extraction using deep learning leverages neural networks to automatically learn and extract features from raw data. This is the scope of Sect. 4.3 and Sect. 5.3.

This Section provides a brief introduction to the usage of handcrafted audio features in MER and will give an overview of some of the handcrafted features used in this thesis.

Audio Features

When describing audio features in music, one must be aware that we describe it on different levels. While emotional content can be seen as a high-level feature, most features used in MER are referred to as low-level features or Low-Level Audio Descriptor (LLD). Additionally, several authors discuss the notion of mid-level perceptual features for characterizing music recordings. These features represent higher-level abstractions that are more perceptually meaningful to humans [8].

Examples of these different levels are:

- High-level: This song is calming and happy.
- Mid-level: This song is written in a major key.
- Low-level: This song has a low zero-crossing rate.

Audio features related to emotions can be divided into rhythmic, timbre, melodic and harmonic features.

Rhythm is a fundamental component of music and has a significant impact on its emotional expression [43]. Rhythmic features capture temporal patterns and regularities in the music signal. Examples of rhythmic features include tempo, rhythmic complexity, rhythmic stability or note duration statistics [43]. Looking back at the spectrogram, rhythmic features can approximately be regarded as vertical lines [28].

Timbre refers to the quality or colour of sound that enables us to distinguish different musical instruments or voices. Timbre features are closely related to the emotional perception of music, as different timbral qualities can evoke distinct emotional responses. They provide information about the harmonic and inharmonic components, as well as the envelope of the sound. A rounder amplitude envelope, for example, is related to negative emotions such as disgust, sadness or fear. At the same time, a sharper one is related to positive emotions such as happiness or surprise [43]. Examples of timbre features include zero-crossing rate, brightness, and spectral flux.

Melodic and harmonic features include pitch, key, modality or register distribution. For example, elements, such as ascending versus descending melodic contours, have been studied and related to several emotions [43]. Looking back at the spectrogram, spectral features can approximately be regarded as vertical aspects [28]. Harmony, in this case, is *how* vertical lines are stacked, and melody is *how* the vertical lines change over time. One commonly used feature are chroma features.

As we described in the previous Section, pitches are preceived as *similar*, if they differ by an octave. "The main idea of chroma features is to aggregate all spectral information that relates to a given pitch class into a single coefficient" [39, p. 128]. The chroma features of the intro of Stevie Wonders *Sir Duke* are vizialized in Fig. 3.4. Let us inspect the first note again. While in the logarithmic spectrogram on Fig. 3.3, we could see many frequencies at the same time, in Fig. 3.4 we can now see that most of those frequencies belong to the pitch class of the B.

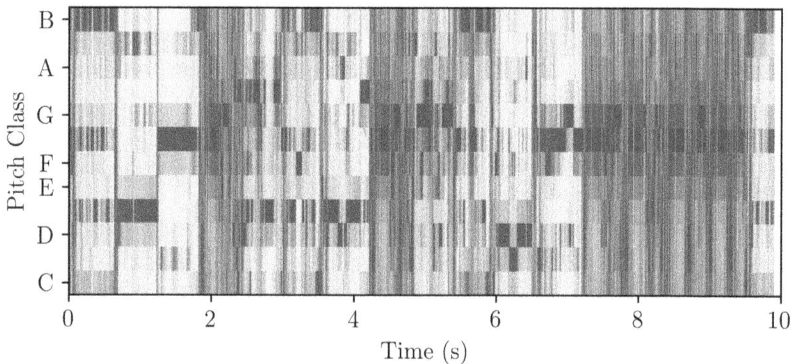

Fig. 3.4 Chroma features of the intro of Stevie Wonder's *Sir Duke*

DEAM Featureset

As part of this thesis, we will use the 260 baseline audio features provided in the DEAM. They are based on 65 LLDs and their first-order derivatives. These 130 LLDs were calculated with a step size of 10 ms [3]. Finally, the authors [16] computed the mean and standard deviation functionals over 1 s time windows with a step size of 0.5 s. This resulted in 260 features extracted at a 2 Hz rate. For the extraction, they used the open-source feature extractor openSMILE [16]. For a more detailed description, the reader is referred to Coutinho et al. [10].

In order to stay within the scope of this thesis, instead of providing descriptions for all 65 features, we will present examples that will give the reader a sense of the features being utilized. For example, there are 13 Mel Frequency Cepstral Coefficients (MFCC). MFCC are an adaption of the time-frequency representation. The idea behind MFCC is to mimic the human auditory system's response to different frequency ranges. The human ear is more sensitive to some frequency ranges than others, and MFCC aim to capture this sensitivity. This enables us to capture the timbral properties of the signal [39]. The **zero crossing rate**, on the other hand, represents the number of times the waveform changes sign in a window (crosses the x-axis). It can be used as a simple indicator of noisiness. For example, due to guitar distortion and heavy percussion, heavy metal music tends to have much higher zero crossing values than classical music [43]. Furthermore, values for the intensities of different filtered versions of the spectrogram are extracted. This represents the acoustic characteristics of a sound signal, particularly its frequency content over time.

Deep Learning 4

In recent years, deep learning has emerged as a powerful and versatile approach to machine learning, enabling breakthroughs in various domains such as computer vision and natural language processing. *In the realm of language generation, transformer-based architectures have ushered in remarkable advancements, allowing us to construct intricate and eloquent sentences, like this one, which effortlessly combines meta humor with the exploration of deep learning's capabilities* [ChatGTP].

This chapter aims to provide the necessary theoretical and mathematical background for understanding deep learning principles, thereby establishing a solid theoretical foundation for the models that will be developed in Chap. 6. Section 4.1 starts with the relevant principles of machine learning, Sect. 4.2 introduces the feed-forward network as the fundamental deep learning architecture and Sect. 4.3 goes into depth of further relevant deep learning architectures and paradigms that build on top of this. Most of the mathematical foundations presented in this chapter are based on the extensive and well-written *Deep Learning Book* by Goodfellow, Bengio and Courville [19].

Before we dive into the theoretical aspects of deep learning, we will clarify some commonly confused terms in this field. *Artificial intelligence*, *deep learning* and *machine learning* are often used interchangeably, which can lead to confusion.

Artificial Intelligence (AI) is a research field with a rich history going back to the 1950s [30]. The goal is to create *intelligent* machines. One of the earliest approaches to this were **rule-based systems**, which were developed in the 1950s and 1960s. These systems work well in environments that can be formally described and have led to a fair amount of success in the first era of AI. An example of a rule-based system is a chess-playing AI. These systems are based on a manually defined symbolic representation of the domain, a set of rules for manipulating these symbols and an algorithm for searching the solution [30]. Since many daily problems

Y. Venohr, *Deep Learning in Personalized Music Emotion Recognition*, BestMasters,
https://doi.org/10.1007/978-3-658-46997-9_4

humans solve cannot be described formally, subsequent AI systems relied on large **knowledge bases** instead of a fixed set of rules [30].

Due to the difficulties faced by systems relying on hard-coded knowledge, rule-based and knowledge-based approaches quickly reached their practical limits, and **Machine Learning (ML)** emerged as a new approach to AI. Since humans acquire knowledge for any task through learning, the focus shifted to algorithms that improved their performance based on data provided to them [30]. ML consists of a wide range of techniques ranging from simple mathematical methods such as SVR or logistic regression to more complex methods such as deep learning. One can say **Deep Learning (DL)** is a particular kind of ML "that achieves great power and flexibility by learning to represent the world as a nested hierarchy of abstract concepts" [19, p. 8].

To sum up, we can say ML is one approach to AI and DL is one particular type of ML technique. Figure 4.1 shows these relations in a Venn diagram.

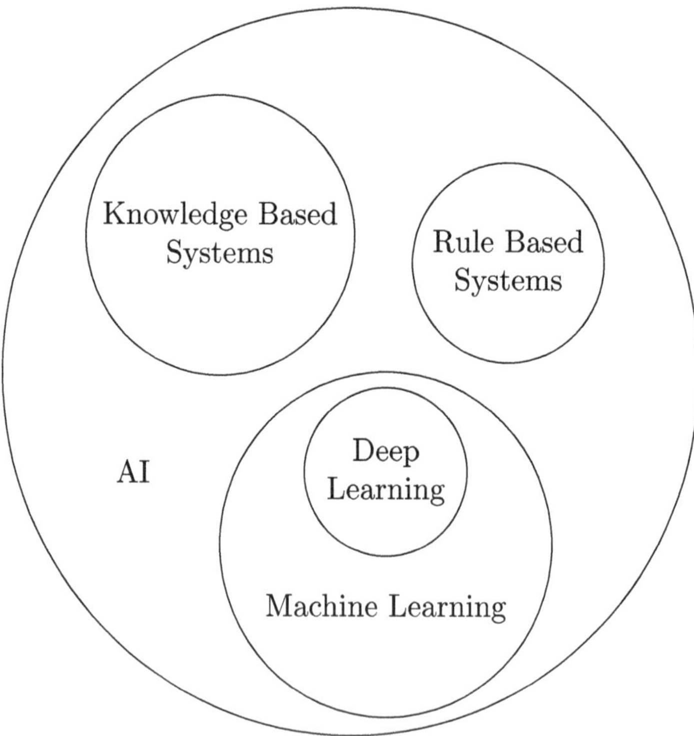

Fig. 4.1 A simplified landscape of artificial intelligence as proposed by [30]

4.1   Machine Learning

Since DL is a specific type of ML, some relevant principles apply to all types of ML. These are in the scope of this chapter.

Let us dive in by defining machine learning:

"A computer program is said to learn from experience E with respect to some class of tasks T and performance measure P, if its performance at tasks in T, as measured by P, improves with experience E" [38, p. 2].

Now we can apply this abstract definition to our specific problem to introduce essential terms and concepts.

Common tasks T for ML include **classification** or **regression** tasks. While a classification task involves determining which of k categories an input belongs to, a regression task involves predicting a numerical value to a given input. The task T in this thesis can be defined as:

Task: *Predict the emotion of a given song.*

Since this thesis uses a dimensional model for emotions (see Chap. 2), the task is a regression task. The output y of our model will be of $\mathbb{R}^2$, where the first value is the valence and the second value is the arousal of a song. Since songs are given as audio files, our input x is a vector of $\mathbb{R}^n$ (see Sect. 3.1). In mathematical terms, our regression task can be seen as a function $f : \mathbb{R}^n \rightarrow \mathbb{R}^2$. So we can put it more precise:

Task: Given a song $x \in \mathbb{R}^n$, predict the emotion $y \in \mathbb{R}^2$.

The experience E a ML algorithm is learning from can be described as experiencing a large collection of examples, usually called the **dataset**. Depending on the experience E a ML algorithm is allowed to have, they can be categorized as supervised and unsupervised learning algorithms.

"Roughly speaking, **unsupervised learning** involves observing several examples of a random vector x and attempting to implicitly or explicitly learn the probability distribution $p(x)$, or some interesting properties of that distribution. In contrast, **supervised learning** involves observing several examples of a random vector x and an associated value or vector y, and learning to predict y from x, usually by estimating $p(x \mid y)$" [19, p. 105].

Even though many ML architectures can perform both supervised and unsupervised [19], we will focus on supervised learning techniques in this section. When introducing the pre-training in Sect. 4.3, we will revisit the concept of unsupervised learning.

A supervised learning algorithm dataset consists of input examples x and their associated target values y. In our case, the examples will be songs and their associated emotion labels from DEAM. We can define it as:

$$D := \{(y_j, x_j) : \forall j \in J_{\text{train}}\},$$

where $x \in \mathbb{R}^n$, $y \in \mathbb{R}^2$ and J is the set of song indices, we use for training.

Even though, regarding our entire pipeline, x is a song's raw audio file, the actual machine learning algorithm usually receives another form of representation of the song that we call **features**. This can be audio features as described in Sect. 3.3 or representations from a music understanding model as described in Sect. 5.3. This more vague form of *some representation of a song* will be denoted as d in this thesis.

The performance measure P is a way of evaluating the abilities of a ML algorithm. In our case, we would like to know how *close* the predicted emotion $\hat{y}$ is to the actual emotion y of the song. In mathematical terms, we would like to have some metric $p : \mathbb{R}^2 \times \mathbb{R}^2 \to \mathbb{R}$ to determine the distance of the predicted and the actual emotion. Obvious choices to calculate this for many examples are Mean Absolute Error (MAE), Mean Square Error (MSE) or Root Mean Square Error (RMSE). We will choose MSE because it is the computationally cheapest [19]. Our performance measure, also called **loss**, can then be defined as

$$\text{loss} := \sum_{j \in J} (y_j - \hat{y}_j)^2.$$

Another way to look at this is to say we have an input x and a target value y, and we assume there is some function f^* that can map these values so that $y = f^*(x)$. But instead of knowing f^*, we just know some examples of x and $f^*(x)$. Our model should define a mapping $y = f(x, \theta)$ and learn the values of the parameter θ that result in the best approximation of f^*.

Generalization

A central challenge in ML is for the model to perform well on inputs the model has not been trained on. The ability to do so is called **generalization**.

Let D_{train} and D_{test} be subsets of D and $D_{\text{test}} \cup D_{\text{train}} = \emptyset$. Now we limit the experience E to D_{train}, to determine the performance P we use D_{test}. While we are training the model, we can calculate the **training error** using our loss function on D_{train}. What we care to optimize is the **generalization error**. The performance of our model on unseen data in D_{test}.

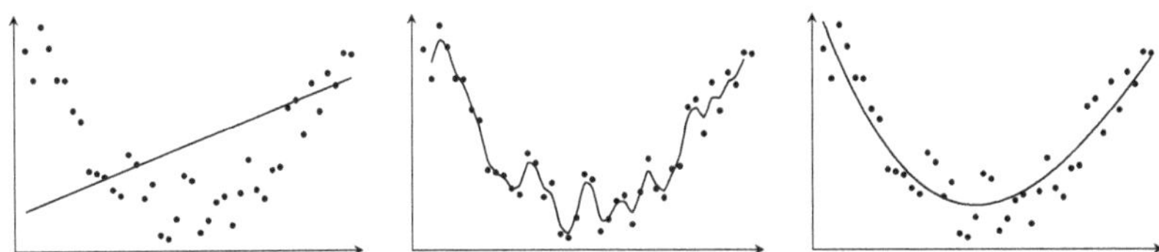

Fig. 4.2 Example for one regression model that is underfitting, one that is overfitting and one that has an appropriate capacity

Now let us consider a simple regression problem, where input and target values are from $\mathbb{R}$ to introduce two important concepts: **underfitting** and **overfitting**. In Fig. 4.2, the results of three simple regression models are shown. On the left, we can see a model underfitting the data. It is not able to detect the curve the data points are making. This will result in a high training error and a high generalization error. In the middle, we can see a model that overfits. We can see that it follows the variance of our sample points. This will result in a low test error, but the overfitting also creates a big gap between the training and the generalization error. One could say that it is "memorizing properties of the training set that do not serve them well on the test set" [19, p. 111].

So in model development, we have two goals:

- Drive the training error as low as possible.
- Try to keep the gap between training and generalization error as small as possible.

A way to control whether a model over- or underfits is to control its **representational capacity**. The capacity describes the model's ability to fit a wide variety of functions. For example, a linear regression that only includes polynomials of degree one has a lower capacity than a linear regression that includes polynomials up to a degree of five. In Fig. 4.2, the model on the left has a low capacity, and the model on the right has a high capacity.

Typically, the generalization error has a U-shaped curve as a function of model capacity [19]. An increase in the model's capacity decreases the generalization error initially, then reaches an optimal point and increases again with a higher capacity. So in order to minimize the generalization error, the goal is to find a model with the optimum capacity for a particular problem. This is closely linked to the idea of the as the **bias-variance trade-off** [23]. While an increase in model capacity decreases our model's bias error, it increases our model's variance error.

4.2   Feedforward Networks

Since most conventional ML methods require careful engineering and a lot of domain expertise to design a suitable internal representation, a set of methods called representation learning has increased in popularity [33]. Deep Learning is a representation learning methods with many layers of representation (hence the name *deep*). The quintessential models for deep learning applications are **feedforward neural networks**. The scope of this section is to describe the mathematics behind training a neural network and introduce important methods and terms. All other models, which will be described in Sect. 4.3 such as convolutional neural networks and recurrent neural networks, are modified versions of it. So in order to understand them, we first need to understand the basic principle of feedforward networks.

Layers of Abstraction

Neural networks are called networks because they are composed of many different chained functions. Each function is a simple but non-linear module that transforms the representation of the raw data to a slightly more abstract level [33]. Given three functions $f^{(1)}$, $f^{(2)}$ and $f^{(3)}$ for instance, our network could look like this

$$f(x) = f^{(3)}\Big(f^{(2)}\big(f^{(1)}(x)\big)\Big).$$

In this case $f^{(1)}$ is called the **first layer** of our network $f^{(2)}$ is called the **second layer** and so on. The last layer is also called the **output layer**. The **depth** of the model is defined by the length of the chain, respectively, the number of layers. In this case, our model has a depth of three layers. The layers in between the input and output layers are called **hidden layers**. The hidden layers are usually vector-valued. The dimensionality of the hidden layers determines the **width** of the model.

Another important term when discussing neural networks is the term **unit**. There are two ways of thinking about the network. Each layer can be seen as a function from vector to vector, but one can also interpret a layer as many parallel units acting as functions from vector to scalar. This idea of using many layers of vector-valued representation is drawn from neuroscience. Hence the name *neural* network. In each unit, a weighted sum of the inputs from the previous layer is computed, and the results are then passed through a non-linear function.

Let us use an example to clarify what this means. Figure 4.3 gives an example of a simple neural net with five layers $a^0, a^1, ..., a^4$. In this example, our three hidden layers each consist of three units. Moving forward through the network, every unit is connected to every next layer unit. Thus these layers are called **Fully Connected (FC)**. A FC layer can be described as

$$y = g(Wx + c),$$

where x is the input vector y is the output vector, W are the **weights** of a linear transformation, c are the **biases** and g is the non-linear **activation function**.

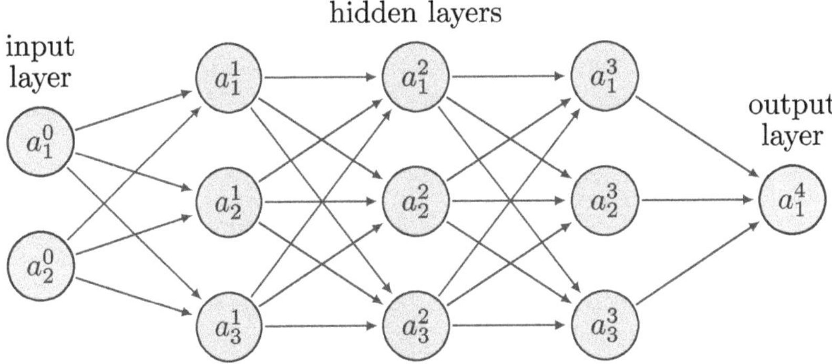

Fig. 4.3 Example for a neural network with three hidden layers

Let us look at our example's first layer a^1 in Fig. 4.3. The input x for this layer is $(a_1^0\ a_2^0)^\mathsf{T}$ and the output y is $(a_1^1\ a_2^1\ a_3^1)^\mathsf{T}$. Since $x \in \mathbb{R}^2$ and $y \in \mathbb{R}^3$, our layer can be described as a function $f^{(1)} : \mathbb{R}^2 \mapsto \mathbb{R}^3$. The weights are represented by the matrix $W \in \mathbb{R}^{2\times3}$, and our biases are represented by the vector $c \in \mathbb{R}^3$. From the perspective of one unit, it can be described as

$$a_1^i = g(x^\mathsf{T} W_{i,:} + c_i).$$

The most popular choice for the activation function g is the Rectified Linear Unit (ReLU) defined as $g(z) = \max(z, 0)$. In the past neural nets used smoother non-linearities, such as $g(z) = \tanh(z)$ or the sigmoid function $g(z) = \frac{1}{1+e^{-z}}$, but it has been shown that ReLU learns much faster in architectures with many layers [33]. Even though ReLU is not differentiable at $z = 0$, this turns out not a problem in practice [19].

Concluding our example, the scalar value for a_1^1, for instance can be calculated by evaluating

$$a_1^1 = \max(W_{1,1}a_0^1 + W_{1,2}a_0^2 + c_1, 0).$$

The next layers use the output of the previous layers and proceed with the same logic, while each layer has its own weights and biases.

Training a Neural Network

Let us say we have a neural net described by f. Looking back to Sect. 4.1, the goal for our model is to learn the values of the parameter θ that result in the best approximation of some function f^*. When talking about neural networks, the values of our parameter θ are represented by the weights and biases of the layers in f. The process of learning the best values for θ is called **training** the model. In the next section, we will look at how to find proper internal parameters for a good approximation of f^*. Since we do not know f^*, we use our labelled examples x and y, which we introduced as training data for this.

Let us start with an overview of the learning procedure:

1. Set (random) initial parameters for θ.
2. Calculate an output $f(x) = \hat{y}$ for a given input x from the training data using the procedure described above.
3. Calculate the loss between the output $\hat{y}$ of the model and the expected value y for this θ. This is called the cost function $J(\theta)$.
4. Calculate the gradient $\nabla_\theta J(\theta)$.
5. Use a gradient-based optimizer to adapt θ in way that the error decreases.
6. Repeat this for N epochs.

Using the Gradient

In order to find the best approximation of f^*, we need to find a θ that minimizes our cost function $J(\theta)$. Starting from the initial (random) value for θ, during training, the value should be adjusted in a way that $J(\theta)$ is the lowest. To adjust θ in a meaningful way, a learning algorithm computes a gradient vector that, for each weight, indicates how strongly the error would increase or decrease if the weight were increased by a tiny amount [33]. This process is called **gradient-based learning**. We can use the gradient to decent towards the point with the lowest costs, respectively, towards the most appropriate θ.

Since gradient-based learning needs the gradient, the next part presents methods to compute the gradient in a deep learning network. In feedforward neural networks, information flows forward through the network. Our model receives x as an input, passes this through the hidden layers and produces $\hat{y}$ as an output. This is called **forward propagation**. Using training data, we can then calculate the cost $J(\theta)$ of our output. The algorithm that then computes the gradient is called **back-propagation** (often simply backprop). The approach is nothing more than a practical application of the chain rule for derivatives, but "despite its simplicity, the solution was not widely understood until the mid-1980s" [33, p. 5]. The gradient of $J(\theta)$ with respect to θ is called $\nabla_\theta J(\theta)$.

Let $x \in \mathbb{R}$ and $f, g : \mathbb{R} \mapsto \mathbb{R}$. If $y = g(x)$ and $z = f\big(g(x)\big) = f(y)$, the chain rule states

$$\frac{dz}{dx} = \frac{dz}{dy} \frac{dy}{dx}.$$

In the case of vectors $x \in \mathbb{R}^m$, $y \in \mathbb{R}^n$ and $g : \mathbb{R}^m \mapsto \mathbb{R}^n$, $f : \mathbb{R}^n \mapsto \mathbb{R}$. If $y = g(x)$ and $z = f(y)$, the chain rule will be

$$\frac{\partial z}{\partial x_i} = \sum_j \frac{\partial z}{\partial y_i} \frac{\partial y_i}{\partial x_i}.$$

In vector notation, this can equivalently be written as

$$\nabla_x z = \left(\frac{\partial y}{\partial x}\right)^{\mathsf{T}} \nabla_y z,$$

where $\frac{\partial y}{\partial x}$ is the $n \times m$ Jacobian matrix of g.

The back-propagation algorithm performs such a Jacobian-gradient product for each layer in our neural net. Conceptually this works the same way with tensors instead of vectors. "We could imagine flattening each tensor into a vector before we run back-propagation, computing a vector-valued gradient, and then reshaping the gradient back into a tensor" [19, p. 207].

Having calculated the gradient, we can use **gradient-based optimizer** to adjust θ and minimize the loss. While simpler ML models like support vector machine or logistic regression have convex loss functions, the loss function of neural networks is non-convex due to the non-linearity in our model [33]. Meaning we do not have a gradient-based optimizer with a global convergence guarantee. We instead use "iterative, gradient-based optimizers that merely drive the cost function to a very low value" [19, p. 177].

There are several approaches to gradient-based optimization. They all use a parameter called **learning rate**, which determines the step size at which the weights θ are being adjusted. A popular choice is ADAptive Moment estimation (ADAM) [31]. ADAM combines the advantages of two other optimization algorithms, namely AdaGrad and RMSProp and has been shown to converge faster and achieve better results compared to other gradient-based optimizers in many practical scenarios [19]. "The method computes individual adaptive learning rates for different parameters from estimates of first and second moments of the gradients" [31, p. 1].

Regularization

In the previous section, we explained the concept of capacity and introduced controlling the capacity as one way of reducing the generalization error. However, there are more methods are intended to reduce the generalization error while not reducing the training error. These methods are called **regularization** methods. Many different regularization techniques have been developed. In fact, some scholars argue that "developing more effective regularization strategies has been one of the major research efforts in the field of ML" [19, p. 228].

A straightforward approach is **multi-task learning**. Let us say we have different tasks, and we want to predict variables y_1, y_2, ..., y_n that share the same input x and share some underlying relation across these tasks. It has been shown that higher generalization can be achieved when training multiple variables simultaneously because of internally shared parameters [19].

When training models with sufficient representational capacity to overfit the task, we often observe that training error decreases steadily over time, but generalization error begins to rise again [19]. A simple but effective technique to prevent this is **early stopping**. The idea is to stop the training process early based on a specific criterion. The criterion could be that the generalization error increased for a fixed number of iterations.

Another technique is called **dropout**. "The key idea is to randomly drop units (along with their connections) from the neural network during training.[...] This significantly reduces overfitting and gives major improvements over other regularization methods" [40, p. 1930]. The so-called **dropout rate** is a value between 0 and 1, determining what fraction of the values are being dropped.

Commonly another technique called **minibatch learning** is applied. This means that the procedure described above is not done on the entire training data but on subsets called a minibatch. The number of examples per minibatch is called the **batch size**. There are different motivations for doing this. One is a significant gain in computational efficiency when using parallel computing, and the other is a regularization effect [19]. The latter leads to a better generalization capability (see Sect. 4.1) and could be explained by the added noise due to smaller batches. Additionally, training with a smaller batch size might require a lower learning rate to maintain stability due to the added noise in the gradient estimate [19]. Every time we update θ for one minibatch, we complete one **iteration**. When our model has iterated through all minibatches, we completed one **epoch**. Then the training is repeated until the model reaches a sufficient performance.

4.3   Deep Learning Architectures

Based on the basic principle from feedforward neural networks, many modifications and adaptions exist for specific learning tasks. This section will present some commonly used architectures for deep learning models.

Convolutional Neural Networks

One common type of architecture widely used is the Convolutional Neural Network (CNN) architecture. It was first introduced and popularized with successfully recognizing handwritten digits in 1989 [58]. CNN are neural networks that use at least one convolution layer (instead of a FC layer) as one of its layers [30]. This architecture has proven successful with input data with a grid-like topology (images as multiple 2D arrays or time series as 1D arrays) [30, 33]. There are four key ideas behind CNNs: shared weights, local connections, pooling and many layers. [33].

Let us first take a look at **convolution layers**. A convolution layer is a layer that uses the **convolution operation**. In its most general form, the convolution operation is defined as the integral of the product of the two functions. In practical applications in deep learning, we work with discrete data in even intervals, like music that is stored in a fixed sampling rate or images that have pixels with equal width. Thus we can define the convolution operation in one dimension using a sum as

$$s(t) = \sum_{a=\infty}^{\infty} x(a)w(t - a),$$

typically denoted as

$$s(t) - (x * w)(t).$$

Usually, x is referred to as the **input**, w is referred to as the **kernel** and the output is referred to as the **feature map**. Conceptually this works the same in higher dimensions, with one sum going along each dimension. As described above, a FC layer can be described as $y = g(Wx + c))$, where x is the input vector, y is the output vector, w is the weights and g is the activation function. A convolution layer can be described as $y = g(s((x * w)))$, where s is the convolution operation.

Discrete convolution can also be viewed as multiplication by a matrix with several entries constrained to be equal to others. These matrices correspond with very sparse matrices (entries are mostly equal to zero). This has the effect that not all input units are connected to all output units. This is called **sparse connectivity** between the layers. Figure 4.4 shows an example of this. Typically the kernel is much smaller than the input. Meaning it has much more zero entries. "For example, when processing an image, the input image might have thousands or millions of pixels, but we can detect small, meaningful features such as edges with kernels that occupy only tens or hundreds of pixels" [19, p. 335].

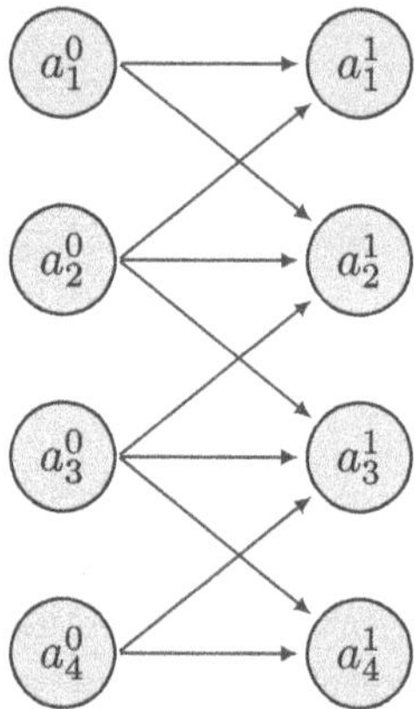

Fig. 4.4 Example for two layers with sparse connectivity, instead of being fully connected (Fig. 4.3). In this case, convolution with a kernel of width three is used. Therefore the input units a_i^0 have a maximum of three connections

Another effect of using the convolution operation is the **parameter sharing**. In a FC layer (Fig. 4.3), each weight is used precisely once to connect one unit to another. Each connection has its parameter. If a layer uses a convolution operation, some parameters in the matrix representing the kernel are shared among different units. In Fig. 4.5, we have the example of a kernel with three elements. The red arrows show all the connections that use the central element of the three elements kernel. Thus all units share this parameter. This leads to the interesting effect that our layer becomes **equivariant** to translation. If we work with furrier transformed audio data (see 3.2), "using convolution in the time makes the model equivariant to shifts in time. Using convolution across the frequency axis makes the model equivariant to frequency so that the same melody played in a different octave produces the same representation but at a different height in the network's output" [19, p. 361]. It is important to note that convolution is not naturally equivariant to other transformations, such as rotations or scale changes.

Another concept that goes hand in hand with using the convolution function in a CNN is using a **pooling function**. Usually, a layer in a convolutional network consists of multiple stages. The first stage performs multiple convolutions in parallel

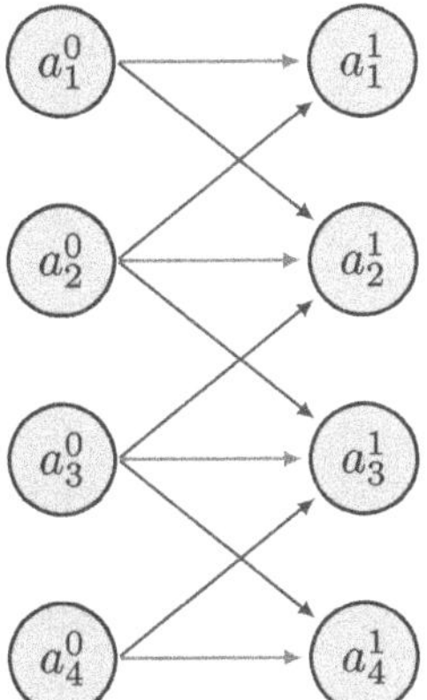

Fig. 4.5 Parameter sharing. Example for two layers with sparse connectivity due to a kernel with three elements. The red arrows indicate that all units use the central element of the three-element kernel. Meaning they all share a parameter

convolutions, the second stage with a non-linear activation (usually ReLU) and the third stage uses a pooling function. The purpose of a pooling function is to replace the output at a certain location with a summary statistic of the nearby outputs. Examples of pooling operations can be using the maximum value or using the average value in a rectangular neighbourhood. The layer becomes approximately **invariant** to small translations using these summarising functions. This means that a small translation of the input results in a very similar pooled output, where most values stay the same.

Having convolutional layers makes sense in array data since local groups of values are often highly correlated, forming local motifs that are easily detected and invariant in location. "In other words, if a motif can appear in one part of the image [or audio], it could appear anywhere, hence the idea of units at different locations sharing the same weights and detecting the same pattern in different parts of the array" [33, p. 4].

Recurrent Neural Networks

For tasks with sequential input, such as language or audio, Recurrent Neural Network (RNN) architectures have proven very successful [22, 33]. This architecture processes the input sequence one element at a time. At the same time, a hidden state implicitly hold information about the history of all the previous elements of the sequence [33].

Let us take one step back first. When talking about RNNs, it makes sense to talk about our input data as sequences $x^{(1)}, \ldots, x^{(\tau)}$ with length τ, where $x^{(t)}$ is a vector representing features at time step t.

To introduce the *recurrence* in our RNN, we can define the hidden state $h^{(t)}$ at time t in a recurrent way:

$$h^{(t)} = f(h^{(t-1)}, x^{(t)}, \theta).$$

$h^{(t)}$ depends on the value of $h^{(t-1)}$, which itself uses the same definition, depending on the values of $h^{(t-2)}$ and so on. For a finite number of time steps, this can be expressed in a way that does not involve recurrence. For example $h^{(3)}$ can be expressed as

$$h^{(3)} = f(h^{(2)}, x^{(3)}, \theta)$$
$$h^{(3)} = f\big(f(h^{(1)}, x^{(2)}, \theta), x^{(3)}, \theta\big).$$

This process is called **unfolding** the computational graph. Figure 4.6 visualizes the unfolding process. The recurrence in the definition has the effect that the parameters weights are shared across all time steps because they all use the same function f. Since only one model f needs to be learned, RNNs becomes independent of the length of our input sequence.

With the RNN we have described, the state at time t only captures information from the past $x^{(1)}, \ldots, x^{(t-1)}$. However, for some tasks the state at time t does not only depend on the previous time steps, but also on the following time steps. For this, Schuster et al. [49] introduced **bidirectional recurrent neural networks**. As the name suggests, bidirectional RNNs "combine an RNN that moves forward through time beginning from the start of the sequence with another RNN that moves backward through time beginning from the end of the sequence" [19, p. 395].

RNNs have been successfully used "to model both sequential inputs and sequential outputs as well as mappings between single data points and sequences" [37, p. 24]. For doing a so-called **sequence-to-one** modelling, there are different approaches. Either the hidden state of the last time step h_t [7] or the average over all time steps [15] is used as the model's output.

Even though RNNs can be very powerful, they have been proven to be problematic in training. Because the backpropagated gradients shrink or grow at each time step, they tend to vanish or explode [33]. This means they become so close to zero that the parameters practically do not move during training or become so big that they jump around, and learning becomes unstable. This is called the **vanishing and exploding gradient problem**. This is because the weights are shared for all

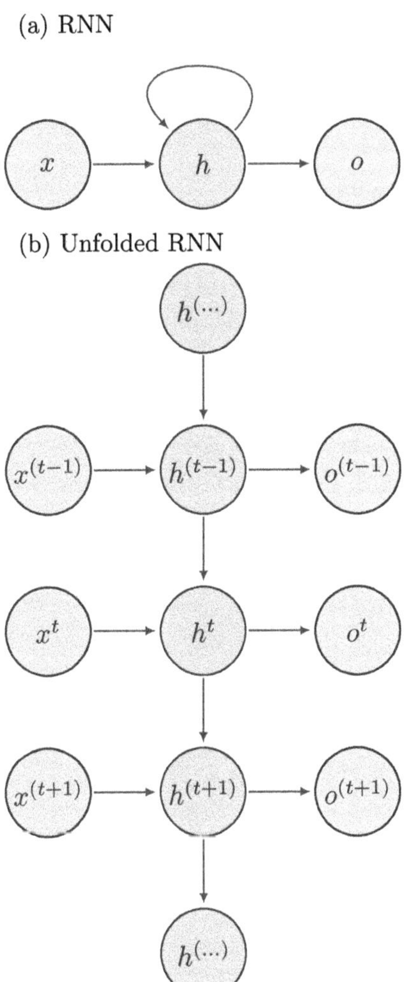

Fig. 4.6 Unfolding a recurrent neural network

time steps. If a weight is 2, for example, the calculation for the gradient includes some expression including 2^{τ}, which can get big quickly. Because feedforward networks do not use the same parameters for each step, "even very deep feedforward networks can largely avoid the vanishing and exploding gradient problem"

[19, p. 290]. Additionally to this problem during training, RNNs have shown difficulties in learning to store information for very long [22, 33]. "The difficulty with long-term dependencies arises from the exponentially smaller weights given to long-term interactions [...] compared to short-term ones" [19, p. 401].

Long Short-Term Memory

One of the most effective approaches proposed to reduce the difficulty of learning long-term dependencies so far is the so-called **gated RNNs**. One example of this is the **Long Short-Term Memory (LSTM)** architecture, which was introduced in 1997 by Hochreiter et al. [27]. This architecture uses Long Short-Term Memory (LSTM) cells with an internal recurrence on top of the outer recurrence of RNNs. These cells have the same inputs and outputs as the previously described RNNs but use an additional system of gating units that control the flow of information. These units are the **cell state unit**, the **forget gate**, the **input gate** and the **output gate**. Mathematically, LSTM can be described as (from Pytorch documentation and Sak et al. [46]):

$$
\begin{aligned}
i_t &= \sigma(W_{ii}x_t + b_{ii} + W_{hi}h_{t-1} + b_{hi}) \\
f_t &= \sigma(W_{if}x_t + b_{if} + W_{hf}h_{t-1} + b_{hf}) \\
g_t &= \tanh(W_{ig}x_t + b_{ig} + W_{hg}h_{t-1} + b_{hg}) \\
o_t &= \sigma(W_{io}x_t + b_{io} + W_{ho}h_{t-1} + b_{ho}) \\
c_t &= f_t \odot c_{t-1} + i_t \odot g_t \\
h_t &= o_t \odot \tanh(c_t)
\end{aligned}
$$

where c_t is the cell state at time t, x_t is the input at time t, and i_t, f_t, g_t, o_t are the input, forget, cell, and output gates. σ is the sigmoid function, $\odot$ is the Hadamard product, W are the weights and b are the biases of the network.

Even though it is quite hard to follow the flow of information in this definition, we can see the recurrences of h_t. It depends on states of previous time steps such as h_{t-1} and c_{t-1}.

Above we described a LSTM that consists of one layer. However, it is common to use LSTM with multiple layers. In this case, the input $x_t^{(l)}$ of the l-th layer is the hidden state $h_t^{(l-1)}$ of the previous layer.

The concept of bidirectionality and LSTM can work together. This architecture is then sometimes referred to as Bidirectional Long Short-Term Memory (BI-LSTM).

Transformer

For some time, LSTM networks have been the state-of-the-art approach to sequence modelling. Even though, especially in natural language processing, they still perform poorly when sentences are too long. In 2017 the **transformer** architecture was introduced in the influential paper *Attention is all you need* by Vaswani et al. [52]. Originally it was introduced for translation tasks, but now it is being used in many different areas, including several audio tasks.

The transformer architecture revolutionized the field by eliminating the need for recurrent connections and simply relying on the **self-attention mechanism**. At its core, the transformer consists of an encoder-decoder structure. The **encoder** processes the input sequence and generates a representation called the **contextualized encoding** by employing self-attention. Self-attention allows the encoder to attend to different parts of the input sequence, capturing dependencies and relationships between words. The **decoder**, on the other hand, takes the contextualized encoding and generates the output sequence by attending to the relevant parts of the encoding using self-attention as well.

Self-Supervised Pre-Training

In the previous sections, we focused on supervised learning. We described models that are trained using datasets that include input and target values. However, the performance of supervised learning is inherently limited by the size and scope of labeled music datasets [20]. We now introduce self-supervised pre-training: a paradigm based on the concept of transfer learning.

> "In the original Transformer paper, the translation model was trained from scratch on a large corpus of sentence pairs in various languages. However, in many practical applications of NLP [natural language processing] we do not have access to large amounts of labeled text data to train our models on. A final piece was missing to get the transformer revolution started: **transfer learning**" [51, p. 6].

Transfer learning means that a model learns with one task and then transfers this knowledge to another task, that is referred to as the **downstream task**. Using the paradigm of self-supervised pre-training, the final process consists of two steps:

1. Pre-Training a model on a large corpus of unlabeled data using self-supervised learning on a different task. This model will serve as the **foundation model**.
2. Adding a regression layer and training the model in a supervised environment for a specific downstream task (e. g. MER) using a smaller labeled dataset. This will be referred to as the **regression head**.

As described in Sect. 4.1 supervised learning requires a labeled dataset like $\{(x_j, y_j) : j = 1, \ldots, N\}$, where N is the number of samples and x_j and y_j

are the corresponding input and label pairs of the j-th sample. With this we were training a model to make predictions $f(x, \theta) = \hat{y}$ for each sample. **Self-Supervised Learning (SSL)** is a special type of unsupervised learning. In SSL the algorithm learns on unlabeled data as $\{(x_j) : j = 1, \ldots, N\}$. The idea is to derive a pseudo input x'_j and pseudo label y'_j for each input x_i and then train to minimize the loss between $f(x'_j, \theta)$ and y'_j. An example of the generation of pseudo labels is **masked-language modeling**. Let $x_j = [x_j^{(1)}, x_j^{(2)}, \ldots, x_j^{(L)}]$ be an input sequence of length L and $M \subset [L]$ is a subset of indices randomly chosen from 1 to L. The pseudo data is defined as

$$x'_j = [\mathbf{1}_{[L]\backslash M}(1) \cdot x_j^{(1)}, \mathbf{1}_{[L]\backslash M}(2) \cdot x_j^{(2)}, \ldots, \mathbf{1}_{[L]\backslash M}(L) \cdot x_j^{(L)}]$$

and

$$y'_j = x_j - x'_j,$$

where $\mathbf{1}_{[L]\backslash M}(x) = 1$, if x is outside the masked indices M and $\mathbf{1}_{[L]\backslash M}(x) = 0$ else [34].

In this self-supervised process of masking and predicting the input, a transformer-based model improves by learning a meaningful representation of the input. It learns to represent the input as a vector in a n-dimensional vector space, where similar inputs are closer to each other. These vector representations are referred to as **embeddings**.

Using these embeddings, we add a regression head that learns to map the output to the desired downstream task such as MER. For this it is common to use FC layers. The regression head is then trained in a supervised environment using labeled data for the desired downstream task.

This chapter aims to consolidate the foundations, that have been given in the previous and provide a comprehensive overview of various approaches to tackle the task of MER. In Sect. 5.1, different classification methods for MER approaches are introduced. Section 5.2 presents a selection of notable research studies in the field of MER and Sect. 5.3 focuses on the pre-trained music understanding model MERT.

5.1 Classifying Approaches

There are many different approaches to the MER task. As well as many different research frameworks that attempt to classify these approaches. A way that has been put forward by Han et al. [22] is to divide MER research into three stages:

- **Domain definition**: In this stage, the emotion model and the dataset are chosen. This has been thoroughly discussed in Sect. 2.1.
- **Feature extraction**: In this stage, useful features for the emotion recognition stage are extracted. These can be either handcrafted features such as the LLDs introduced in Sect. 3.3 or features extracted from a DL model.
- **Emotion recognition**: This stage does the regression from features to emotion labels. This can be either ML algorithms such as SVR or DL algorithms such as combinations of LSTM or FC layers.

In MER research, a general shift toward using DL at the last two stages has been observed [22]. In order to describe this shift more systematically, we propose a simplified schematic overview of approaches taken to the MER task. This overview and representative work is shown in Table 5.1.

Y. Venohr, *Deep Learning in Personalized Music Emotion Recognition*, BestMasters,
https://doi.org/10.1007/978-3-658-46997-9_5

Table 5.1 Classification of approaches to the MER task

Name	Input/Transform	Features	Emotion Recognition	Work
(a)	raw audio	handcrafted	ML, e.g. SVR	[42, 43, 55]
(b)	raw audio	handcrafted	LSTM + FC	[15, 28]
(c)	time-frequency	CNN	LSTM + FC	[12, 26, 60]
(d)	raw audio	CNN	LSTM + FC	[41]
(e)	raw audio	found. model	FC	[34]

5.2 Approaches to MER

The first approaches to MER used handcrafted feature extraction and combined this with traditional ML models (Table 5.1 a). There are several established libraries for extracting relevant features, such as the *OpenSmile toolkit* [16]. The process of handcrafted feature extraction has been discussed in 3.3. Different classification or regression models are used depending on whether the emotion model is categorical or dimensional. The most used models for categorical MER are support vector machines, k-nearest neighbours, decision trees, random forest, and naive Bayes [22]. For dimensional MER, the most used regression models are support vector regression, linear regression, multivariate linear regression, Gaussian process regression and acoustic emotion Gaussian [22].

One of the earliest works to regard MER as a regression problem from 2008 is by Yang et al. [55]. The authors used different libraries to extract 114 audio features and employed SVR for static emotion recognition. Using this approach they reached R^2 statistics of 0.28 for arousal and 0.58 for valence. One can notice the big difference between the two dimensions. Aljanaik et al. state that "it is a known issue, that valence is much more difficult to model than arousal" [3]. While most subsequent work started utilizing DL for the emotion recognition stage, Panda et al. [42, 43] have focused on choosing the right handcrafted audio features for the MER task using SVR.

Extracting emotionally relevant features does not only require a lot of prior music knowledge; the omission of necessary features or the inclusion of irrelevant features can also lead to undesirable results. Thus researchers have begun to use lower-level representations, including simple time-frequency representations and other LLDs [17]. Simultaneously, due to their ability to capture temporal information, the use of DL methods based on RNN such as LSTM and BI-LSTM have been employed.

Usually, they are combined with FC layers for the final mapping to emotions (Table 5.1 b).

Methods based on RNN were first applied to emotion recognition, not in the domain of music, but in human emotion detection from speech, facial expression and shoulder gesture [3]. While their use was mostly motivated for solving dynamic emotion recognition, Wenninger et al. [15] found that LSTM outperforms support vector regression even for static MER. The so-called Munich LSTM-RNN approach [15] is an approach taken by a research group from TU München in the 2013 MediaEval *Emotion in Music* task. As input for their model, they used a large feature set of LLDs developed for speech emotion recognition. Their input, as well as their emotion labels, were standardized to zero mean and unit variance. As a loss function, they used the sum of the squared errors. As a model, they used a BI-LSTM with two hidden layers and a size of 128 units per layer. They found that "best song level results in terms of R^2 are obtained by averaging BI-LSTM predictions, outperforming SVR by a large margin for valence [...] while being on par for arousal" [15, p. 2].

For the increased performance in the valence dimension when using RNN based architectures, Lio et. al [28] propose an explanation:

"Arousal depends highly on four features, energy of harmonic sound, energy of percussive sound, tempo of harmonic sound, tempo of percussive sound [...]. Valence has a relatively low correlation with these four features but depends more on features extracted by RNN" [28, p. 2].

Furthermore, the authors propose an architecture where they combine constant LLDs and the output of a BI-LSTM and then feed both of them to a FC layer for regression. To extract the constant features, the music piece is separated into harmonic and percussive sounds using a harmonic percussive source separation algorithm and using tempo estimation algorithms, they can get the tempo of the respective sound. As input for their BI-LSTM, they use the chroma spectrum, where each spectrum corresponds to a twelve-dimensional vector, corresponding to twelve tones of an octave degree. Their BI-LSTM layer has 512 units and an input size of 12. Chroma-based and constant features are combined and fed to two FC layers with each 265 units. Additionally, they introduce a label space defined by pairs of antonyms and find that this can better describe emotions in the context of emotion recognition, and it is easier to train models.

While these approaches have shown state-of-the-art performance, according to Zhang et al. [60] there were still difficulties in extracting relevant features. Since CNNs have been successfully used for feature extraction in image recognition, researchers started to use them to extract features from the time-frequency representation of music because of their image-like shape. Because this approach outperformed handcrafted features, researchers attempted to combine the automatic feature extraction of CNN and the sequential information processing of RNN (Table 5.1 c). There are many different names for these combinations of RNN and CNN.

Dong et al. [12] proposed a so-called *bidirectional convolutional recurrent sparse network* by fusing CNN and LSTM units to better learn the sequential information about emotion from the spectrogram and verified the effectiveness of combining CNN and RNN. Hizlisoy et al. [26] not only introduce a new dataset for MER in Turkish traditional music but also propose an approach for MER using what they call *convolutional long short-term memory deep neural network*. Their architecture uses the output of four CNN layers as features for an LSTM layer and two FC layers for classifying.

Also, some researchers utilize more complex CNN configurations that have been proven successful in image recognition for the MER task. Cramer et al. [11] used VGG methods and found that they can improve the performances of the previous baseline models.

Berardinis et al.[29] additionally consider the role of different musical voices in predicting the emotions of music. Thus they first use source separation algorithms for breaking up music signals into independent song elements (vocals, bass, drums, other). For the prediction for each voice, they use a VGG-style architecture on log-mel spectrograms.

Some research has been done on feeding the raw audio data to the DL model instead of a time-frequency representation (Table 5.1 d). Orjesek et al. [41], for example, present an CNN-RNN approach that maps raw audio data into a sequence of valence-arousal values. Their architecture consists of two parts. A short-term audio feature extractor consisting of a one-dimensional convolution layer and an autoencoder-based interactive reconstruction unit. And a back-end consisting of a RNN, which captures temporal variations of the features, and a maxout fully connected layer, that maps to the valence-arousal values. At the same time, the authors state that "such front-end analysis is not optimal in extracting emotion-specific information because it is not built on any solid theory of emotion expression and/or perception" [41].

## 5.3	Music Understanding Models

Inspired by the success of pre-trained language models in various domains (e. g. ChatGTP), the concept of self-supervised pre-training has recently extended to the field of MER (Table 5.1 e). The foundations of this have been introduced in Sect. 4.3. This research direction aims to develop foundation models that acquire an implicit semantic understanding of music. These so-called **music understanding model** should be able to present music as a meaningful vector in an n-dimensional vector space, where *similar* music is close to each other. This vector representation is referred to as the **embedding**.

While there were some approaches to creating music understanding models in the last two years (e.g. [20, 35]), they lacked computational efficiency and generalization abilities and were not made publicly available [34]. In March 2023, a significant milestone was achieved with the release of the first open-source music understanding model called **acoustic Music undERstanding model with large-scale self-supervised Training (MERT)**. In June 2023, the associated paper by Li et al. [34] was made publicly available. It is important to note that as of the day of writing this, it is only available as a preprint and has not been peer-reviewed yet.

MERT has been released as a pre-trained model, including weights on the Hugging Face platform. It has been released as a small version (MERT-95M) that was trained on 1K hours of and as a large version (MERT-330M) that was trained using 160K hours of music recordings. Both models were trained using the masked-language modelling paradigm (see Sect. 4.3) with 5-second audio snippets as context.

The model is trained to use raw audio data with a sampling rate of 24 000 hertz as an input and produce a tensor of shape (1024, 25) at a sampling rate of 75 Hz as an output. Each time step can be interpreted as 25 layers of a 1024-dimensional embedding vector. Each layer performs better for different downstream tasks [34].

On top of potentially having a better performance, music understanding models offer additional benefits. As stated in Sect. 2.3, a big issue in MER research is copyright restrictions that make it impossible to create big publicly available datasets for research. This applies to MER and to many other music-related tasks in the field of MIR.

"First, PLMs [Pre-trained Language Models] can potentially pave the way to unify the modelling of a wide range of music understanding, or so-called MIR tasks including but not limited to music tagging, beat tracking, music transcription, source separation, etc., so that different tasks no longer need detailed models or features. Second, releasing a PLM for acoustic music understanding could re-distribute the musical knowledge rather than the data itself, which saves the high costs of manual annotation and at the same time is not restricted by copyright laws" [34, p. 2].

MERT has been evaluated on 14 different downstream tasks such as music tagging, key detection, genre classification, emotion score regression, instrument classification, pitch classification, vocal technique detection or singer identification. While closely matching state-of-the-art in most, they achieved new state-of-the-art in 4 of them [34]. This makes it an exciting research direction with great potential for advancing the field of MER and related MIR tasks.

Model Development

6

In the scope of this thesis, two MER models based on entirely different architectures have been developed. This chapter describes the data preprocessing, the DL architecture and training pipeline of the proposed models. The two developed models, will be referred to as:

- **LSTM model:** This model uses the handcrafted audio features provided by DEAM and a many-to-one LSTM architecture with a FC layer for regression.
- **MERT model:** This model a uses the audio embeddings of the music understanding model MERT and a CNN + FC architecture as a regression-head.

Section 6.1 discusses the process of data preparation for machine learning and introduces the specific subsets that will be used in our experiments. Section 6.2 describes the architectural design and learning pipeline of our proposed models and Sect. 6.3 shows the process of optimizing the models by finding proper hyperparameters.

Along the way, code snippets of the implementation will be presented. Since these code snippets just serve as a clarification for the reader to follow, they might be simplifications and can differ slightly from the actual code used.

6.1 Preparing Data

Before we can start with the development and training of our models, it is crucial to prepare the data adequately. This section highlights the key aspects of data preparation and presents a comprehensive overview of the data preprocessing pipelines, including code examples.

Y. Venohr, *Deep Learning in Personalized Music Emotion Recognition*, BestMasters,
https://doi.org/10.1007/978-3-658-46997-9_6

Train, Test and Validation Split

Let D be our dataset. We define it as

$$D := \{(d_j, y_j) : \forall j \in J\}$$

where d_j is the feature representation and y_j is the emotion label of the j-th song of the 1744 songs with indices $j = 1, ..., 1744$ from our set of Indices J. y_j is a vector of size (2), where the first value represents valence and the second value represents arousal of the song.

> "Ideally, the model should be evaluated on samples that were not used to build or fine-tune the model so that they provide an unbiased sense of model effectiveness. When a large amount of data is at hand, a set of samples can be set aside to evaluate the final model" [32, p. 67].

Since we have a sufficient amount of data in our dataset, we can separate one part for the final evaluation of our dataset, **the test set** D_{test} with indices J_{test}. The rest of the data will be split into one subset for training the model, **the training set** D_{train}, and one subset for evaluating the model results while developing the model, **the validation set** D_{val}. Canonically we define indices J_{train} and J_{val}.

A common split is to leave 25% for testing, 50% for training and 25% for validation [23]. This is also the split used in this thesis. The split was done on a randomly shuffled list of D, so we do not get an uneven distribution due to some ordering of the songs in DEAM. For reproducibility, we provide a seed before shuffling. Additionally the indices of J_{test}, J_{train} and J_{val} can be found in the repository. Figure 6.1 shows the distribution of emotions of songs in the test set compared to the rest of the dataset.

While splitting the data, we aim to generate homogeneous subsets and a test set that is representative of our statistical population [23]. Alternatively, resampling methods, such as k-fold cross-validation or leave-one-out cross-validation, can be used to produce more appropriate estimates of model performance. Theoretically, this would be the best approach for using the data as efficiently as possible. At the same time, 10-fold cross-validation means training the model 10 times and thus will be computationally too expensive and exceed the scope of this thesis. Due to this reason, we will assume that our sub-sets have a similar distribution of emotions and that our test set is representative of our statistical population.

To gain a little more confidence in our assumption, we can compare the mean $\bar{y}$ and standard deviation σ of our test set and the entire statistical population (all data points).

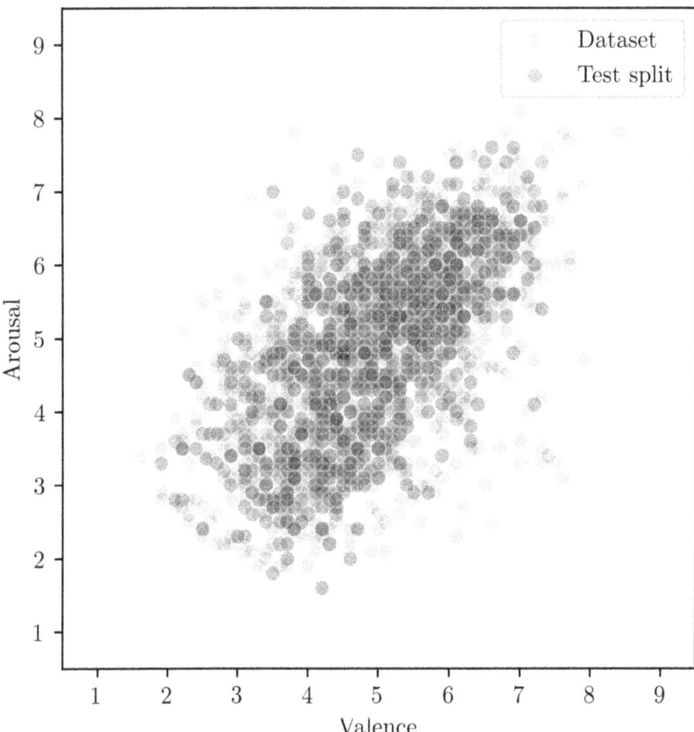

Fig. 6.1 Distribution of test split compared to entire dataset

$$\overline{y}_{\text{test}} = (.546\ .539)$$

$$\overline{y}_{\text{all}} = (.544\ .534)$$

$$\sigma_{\text{test}} = (.122\ .140)$$

$$\sigma_{\text{all}} = (.130\ .143)$$

It is important to note that the values given above are not in the original 9-point rating scale but rather on a normalized scale from 0 to 1. We can see that the values for valence and arousal are similar on average. Even though our test set does have a slightly smaller standard deviation in the valence dimension, this comparison increased the confidence in this assumption.

Dataset Class

Before training a model, we need to load and prepare the data set. Since we ideally want our dataset code to be decoupled from our model training code for better readability and modularity, we will create a dataset class using the data primitive `torch.utils.data.Dataset` provided by Pytorch and call it `DeamDataset`. We can define which split we want and what type of features we want to use on initialization of this class. The class then prepares the annotations y and features d for this split. The preparation and filtering will become more relevant for the experiments conducted in Sect. 7.2.

The heart of our class is the `__getitem__` function, that enables us to use simply use an index on our dataset to retrieve data:

```
def __getitem__(self, index):
    label = self.spsget_label(index)
    features = self.spsget_features(index)
    return features, label
```

Code example 6.1 Function `__getitem__` of the `DeamDataset` class

The aim of this function is to be able to abstract over the detailed steps and build a function that simply gets an index j of a song and returns the pair of (d_j, y_j) for this song. This enables us to simplify methods such as batch learning later on. The function receives an index as an integer as an input and returns the tuple of `signal` and `label`. These are the input and target values our model will use for training. `label` is a two-dimensional vector representation of the emotion label. Since valence and arousal are given on 9 point scale, we normalize them to be valued between 0 and 1. `features` is tensor d_j with the features representing our song j. The exact shape of d_j differs between our two models.

Audio features for the LSTM model

For the LSTM model, we will use the handcrafted features that are provided by DEAM as described in Sect. 3.3. Since they are given as single `.csv` files, we can just load them using the `pandas` library. They consist of 260 features extracted at a rate of 2 Hz for audio of 44 seconds. Thus d_j for our LSTM model is a tensor of shape $(88, 260)$.

Calculating Embeddings with MERT

For our MERT-based model, we use the pre-trained music understanding model MERT to calculate vector embeddings for the songs, which will then be used as features by our regression head. One could also see this as part of our DL-model and see the raw audio as our model input. Then this part can be seen as the foundation model and the rest as the regression head. However, since we are just training the regression head, we will look at the embedding calculation as a preprocessing step.

The pre-trained model MERT was obtained from the Hugging Face platform at https://huggingface.co/m-a-p/MERT-v1-330M. For calculating the embeddings, we adapted the example of the documentation. First we load MERT and its processor:

```
1 mert = AutoModel.from_pretrained("m-a-p/MERT-v1-330M",
      trust_remote_code=True)
2 processor = Wav2Vec2FeatureExtractor.from_pretrained( "m-
      a-p/MERT-v1-330M", trust_remote_code=True)
```

The clue behind `from_pretrained` is that we get the model, including the weights learned in pre-training. Next, we load the raw audio as described in Sect. 3.1, transform the signal to mono and resample it to the sampling rate the processor is expecting and call it `input_audio`. Then, we can pass it through the processor to obtain the inputs for MERT and let the model calculate the embeddings:

```
1 inputs = processor(input_audio, sampling_rate=
      resample_rate, return_tensors="pt")
2 with torch.no_grad():
3     outputs = mert(**inputs, output_hidden_states=True)
```

As described in Sect. 5.3, the outputs are 25 layers of a 1024 dimensional vector at a sampling rate of 75 Hz. Thus for our audio samples, we receive 25 tensors of of shape (3379, 1024). When we stack the 25 layers we have a tensor of shape (25, 3379, 1024). As proposed by the authors in their example on hugging face, and since we are only interested in one label for the entire song, we reduce the time dimension by calculating the mean along the time dimension:

```
1 all_layer_hidden_states = torch.stack(outputs.
      hidden_states).squeeze()
2 time_reduced_hidden_states = all_layer_hidden_states.mean
      (-2)
```

Finally we receive a tensor of size (25, 1024) as an input d_j for our MERT-based model.

The calculation of the embeddings is computationally quite expensive. Since we were using a laptop with an Intel i5 processor, 16 GB of RAM and no GPU, the calculation time for all songs was approximately 30 hours.

6.2   Model and Training Architecture

This Section will describe the DL-architecture of the two proposed models and the training pipeline used in this thesis.

For both models, we are using a multi-task learning approach, meaning that our model will learn to predict valence and arousal simultaneously. The idea behind this is that the model will have a better generalization because the model can have some shared internal representation for both valence and arousal (see Sect. 4.1 on regularization).

LSTM Architecture

As a basis for this model, we will use the LSTM architecture that was introduced in Sect. 4.3. In an analysis of the different approaches taken in the DEAM challenges 2013, 2014 and 2015, the authors find that models based on "fine-tuned LSTM recurrent neural networks are the best-performing models" [3, p. 17] for the MER task. Also, in the years after this challenge, LSTM based models have widely and successfully been used in MER [10, 24, 26, 28, 29, 41, 47, 59, 60].

Our model consists of n LSTM layers, a dropout layer and two FC layers. Since our input is a sequence (song) and our output is one value for the entire song (static MER), we will use a LSTM of the sequence-to-one type. This means that we only use h_t for the last time step t as output from the LSTM layers. For regularization, we add a dropout layer. The fully connected layers handle the mapping to our valence and arousal values.

To implement our model, we will define a class using the data primitive `torch.nn.Module` provided by Pytorch and call it `LSTMNetwork`:

```python
class LSTMNetwork(nn.Module):
    def __init__(self, input_size, hidden_size,
    num_layers, dropout_rate, bidirectional):
        super().__init__()
        self.hidden_size = hidden_size
        self.num_layers = num_layers
        self.lstm = nn.LSTM(input_size, hidden_size,
    num_layers, batch_first=True,
                            bidirectional=bidirectional)
        self.D = 2 if bidirectional else 1
        self.fc_1 = nn.Linear(hidden_size * self.D,
    hidden_size)
        self.fc_2 = nn.Linear(hidden_size , 2)
        self.relu = nn.ReLU()
        self.dropout = nn.Dropout(dropout_rate)

    def forward(self, x):
        batch_size = x.size(0)
        h0 = torch.rand(self.num_layers*self.D,
    batch_size, self.hidden_size)
        c0 = torch.rand(self.num_layers*self.D,
    batch_size, self.hidden_size)
```

```
18      out, _ = self.lstm(x, (h0, c0))
19      out = torch.mean(out, dim=1)
20      out = self.dropout(out)
21      out = self.fc_1(out)
22      out = self.relu(out)
23      out = self.dropout(out)
24      out = self.fc_2(out)
25      return out
```

Code example 6.2 Implementation of the proposed LSTM model using Pytorch

Since we want to experiment with LSTM and Bidirectional Long Short-Term Memory (BI-LSTM), the class is implemented in a way that it can be initialized as either one of them just by passing a boolean called `bidirectional`. For easier notation, we introduce D, which is 1 for a LSTM and 2 for a BI-LSTM. The heart of this class is the `forward` function that defines our forward propagation. It receives a tensor as input x, with the features for our song(s). In the previous Section, we stated that one song is represented as a tensor of shape $(88, 260)$. Since we will use minibatch learning, our input x in this case is a tensor of shape $(N, 88, 260)$, where N is the batch size. We initialize the hidden state h_o and the initial cell state c_0 with random values between 0 and 1. In line 17, the output `out` of the actual LSTM model is generated. It is a tensor of shape $(L, 88, D \cdot H_{out})$, where H_{out} is the hidden size of our LSTM. Since we want to map sequence-to-one, we average all time steps in line 18. This is one of the two techniques we introduced in Sect. 4.3. After initial experiments with only using the hidden state h_t for the last time step $t = 88$ (line 19), this approach was discarded.

The tensor of shape $(N, D \cdot H_{out})$ is sent through a dropout layer and then passed to the two fully connected layers that do the final mapping to our two-dimensional final output. There is one FC layer mapping from $(N, D \cdot H_{out})$ to (N, H_{out}) and one FC layer mapping from (N, H_{out}) to $(N, 2)$, that are connected with a Rectified Linear Unit (ReLU). The decision for the hidden size of the second FC to be equal to H_{out} was motivated for the sake of simplicity and reducing the number of parameters that have to be set.

With 2 layers and a hidden size of 256, this LSTM model has 2 769 666 learnable parameters. This size of θ was determined using the `torchinfo` library. For a detailed description on how to determine the learnable parameters, the reader is referred to Sak et al. [46].

MERT Architecture
Even though the basic setup of our class looks quite similar to the `LSTMNetwork` class we defined previously, the architecture of the regression head of the MERT-based model is a lot simpler.

```
1   class MERTRegressionModel(nn.Module):
2       def __init__(self, hidden_size, dropout_rate):
3           super().__init__()
4           self.aggregator = nn.Conv1d(in_channels=25,
        out_channels=1, kernel_size=1)
5           self.fc_1 = nn.Linear(1024, hidden_size)
6           self.relu = nn.ReLU()
7           self.fc_2 = nn.Linear(hidden_size, 2)
8           self.dropout = nn.Dropout(dropout_rate)
9
10      def forward(self, x):
11          out = self.aggregator(x)
12          out = out.squeeze(1)
13          out = self.dropout(out)
14          out = self.fc_1(out)
15          out = self.relu(out)
16          out = self.dropout(out)
17          out = self.fc_2(out)
18          return out
```

Code example 6.3 Implementation of the proposed MERT-model using Pytorch

In the previous Section, we stated that one song is represented as a tensor of shape (25, 1024). Again, since we are learning in minibatches, our input x is of shape (N, 25, 1024), with N being the number of songs per batch. As proposed in the documentation by Li et al. [34], we are using a CNN with kernel size 1 as an aggregator to map the 25 layers to one. By using the CNN layer, we enable our model to learn which of the 25 layers performs best for the MER task. It learns a weight for each layer and then returns a weighted average of the 25 layers. The resulting tensor of shape (N, 1024) is then passed to the two fully connected layers that do the final mapping to our two-dimensional final output. There is one FC layer mapping from (N, 1024) to (N, H_{out}) and one FC layer mapping from (N, H_{out}) to (N, 2), that are connected with a ReLU.

With a hidden size of 32, this model has just 32 892 learnable parameters.

Training

Now that we defined our two model classes and our data loader class, we can start to train the models. We will exemplify the process using the LSTM-model. The training pipeline for the MERT-based model is analogous.

As a first step, we use our previously defined `DeamDataset` class to load our training set D_{train}:

```
1 from deamDataset import DeamDataset from torch.utils
      .data import
2 DataLoader
3
4 train_dataset = DeamDataset(mode="train")
5 train_dataloader = DataLoader(train_dataset,config["
      batch_size"])
```

Typically, we want to pass our samples in minibatches (see Sect. 4.2) with a defined batch size. For this, we use pytorch's `DataLoader` class to wrap the `train_dataset` instance. After this, our `train_dataloader` is iterable that returns a batch of data on each iteration.

Next, we initialize our previously defined `LSTMNetwork` class with our desired configurations:

```
1 model = LSTMNetwork(
2     input_size=INPUT_SIZE,
3     hidden_size=config["hidden_size"],
4     num_layers=config["num_layers"],
5     bidirectional=config["bi"],
6     dropout_rate=config["dropout_rate"])
7 )
```

Looking back at Sect. 4.1, we also need to define a **loss function** that will be optimized during training. As stated earlier, we use MSE as a loss function. Furthermore, we need to choose an algorithm that uses the result of the back-propagation and optimizes our model. We will choose ADAM algorithm since it has successfully been used by several other researchers that trained LSTMs for MER[24, 26, 41, 47, 60]. The implementation with PyTorch is straightforward:

```
1 loss_fn = torch.nn.MSELoss()
2 optimizer = torch.optim.Adam(model.parameters(), lr=
      config["lr"])
```

Code example 6.4 Define loss function and optimizer

Now that we initiated our `model`, our `data_loader`, our `loss_fn` and our `optimizer`, we can start training. For this, we first define a function that will train one epoch:

```
1 def train_single_epoch(model, data_loader, loss_fn,
      optimizer, device):
2     for input, target in data_loader:
3         prediction = model(input)
4         loss = loss_fn(prediction, target)
```

```
5            optimizer.zero_grad()
6            loss.backward()
7            optimizer.step()
8        return loss.item()
```

Code example 6.5 Train single epoch

The `for`-loop iterates through our minibatches and does the following steps for each batch:

1. Line 3: Make a prediction based on the input with the current model weights.
2. Line 4: Calculate the loss for this prediction.
3. Line 5: Reset any previously calculated gradients.
4. Line 6: Use the backprop algorithm to calculate the gradient that minimized this loss.
5. Line 7: Use ADAM to adjusts the weights according to the gradient.

After training one epoch, we return the loss of our final batch. We will later refer to this as the **training loss**. Looking back at the six steps of the learning procedure that was introduced in Sect. 4.2, the steps described here are the implementation of steps two to five.

Now that we defined `train_single_epoch`, we can repeat this pattern for as many epochs as necessary. For this, we wrap the function in another `for`-loop:

```
1 def train_model(config):
2     for epoch in range(config["epochs"]):
3         train_loss = train_single_epoch(model,
4     train_dataloader, loss_fn, optimizer, device)
5         val_loss = validate_single_epoch(model,
6     validate_dataloader, loss_fn, device)
7         session.report({"epochs": epoch, "train_loss":
8     train_loss, "val_loss": val_loss})
```

Code example 6.6 Train multiple epochs

After each epoch is trained, we use the whole validation set D_{val} to calculate the **validation loss** of our model at this point. In Sect. 6.3, we will observe how training and validation loss change over the course of some epochs to gain insight into how our model is behaving. That is why we report these values.

In this final `train_model` function, we have everything neatly wrapped to train our model and get feedback on the performance. The function receives a variable called `config`, a dictionary containing all the relevant parameters that must be set.

Hyperarameters

Considering the proposed pipeline and model architecture, some parameters must be set before training the model. These parameters are usually referred to as **hyperparameters**. Hyperparameters are all parameters set before the training begins and remain constant throughout the training phase. Unlike the model parameters θ, which are learned from the data, hyperparameters control the model's learning process and architectural choices.

An overview of our hyperparameters and a brief description are given in Table 6.1. To gain a better overview, we will categorize them into parameters that affect the model architecture and parameters that affect the learning process. Additionally, Table 6.2 suggests some initial values for these parameters derived from literature that was presented in Chap. 5. This was only done for the LSTM-model because we could not find literature that used a similar architecture to the MERT-model and mentioned their hyperparameters.

Table 6.1 Summary over all hyperparameters

Category	Parameter Name	Description
LSTM	Input Size	Input size of the LSTM layer. Equal to a number of features.
	Bidirectional	Whether to use a LSTM or a BI-LSTM architecture.
	LSTM Layers	Number of LSTM layers of the model.
	LSTM Hidden Size	Hidden size of each LSTM layer.
MERT	FC Hidden Size	Size of the last FC layer.
Learning	Batch Size	Number of examples in one batch.
	Epochs	Number of complete passes through the training dataset.
	Learning Rate	Learning rate of the optimization algorithm.
	Dropout Rate	Fraction of units set to 0 each iteration.

Table 6.2 Initial values for LSTM model hyperparameters derived from literature

Category	Parameter Name	Initial Value
LSTM	LSTM Layers	3 [9]; 2 [26]; 2 − 4 [59]
	LSTM Hidden Size	512 [24, 28]; 200 [9, 26]
	FC Hidden Size	512 [24]; 200 [9, 26] 265 [28], 128 [47]
Learning	Batch Size	10 [26]; 32 [41]; 4 [47]
	Epochs	≈ 25[24]; 100 [26]; 50 [59]
	Learning Rate	$5 \cdot 10^{-6}$ [9]; 0.0001 [26]; 0.005 [59]; 0.001 (decreasing 1/10 every 10 epochs) [60]
	Dropout Rate	0.5 [24, 59]; 0.25 [41]; 0.8 [47]

6.3   Optimizing the model

To help define the scope of this Section, we will start by highlighting the distinction between model development and model assessment/evaluation phases:

- **Model development:** We want to estimate the performance of different model configurations or different sets of hyperparameters in order to choose an optimal one. We do this by evaluating them on the validation set (see 6.1). During model development, the test set stays locked away. This is the scope of this section.
- **Model assessment/evaluation:** Having chosen a final model, we want to estimate its performance (generalization error) on the test set. This will be the scope of Chap. 7.

Before we dive into the optimization of our model, we will conduct some simple experiments to show some of the theoretical concepts introduced in Sect. 4.1.

One concept we introduced is the concept of representational capacity. We know we can control our model's capacity by controlling its hyperparameters. A larger hidden size for neural networks leads to a higher representational capacity. Let us train the MERT model with a large hidden size of 512, thus with a high capacity and with a dropout rate of 0, thus no regularization.

To gain more insight into our model's performance, we will use the training loss (training error) and the validation loss (generalization error) reported on each epoch by our training function. In Fig. 6.2, we can see training and validation loss on the y-axis and the epochs on the x-axis. We can observe that the training error keeps decreasing each epoch, while our generalization error (validation error) stays in the same range and worsens toward the end. This means the model is getting better at predicting the emotion of the songs that it is training on while not getting better at predicting the emotion of the songs in our validation set. As soon as we can observe this, our model is not learning any underlying logic we want our model to learn, but it is starting to learn the variance of the data. In this case, the model has a representational capacity that is too high and thus is overfitting.

In order to prevent overfitting, we implemented several regularization techniques in our model. One such technique was early stopping, where we introduced a counter in the `train_model` function. This counter increments if the validation loss exceeds the previous loss. Once the counter reaches a predefined threshold, we stop the training process, indicating that the model's performance is no longer improving.

After 30 epochs, our model achieved a MSE of 0.01 on the validation set. However, upon analyzing the curve in Fig. 6.2, we observed that the lowest point on the

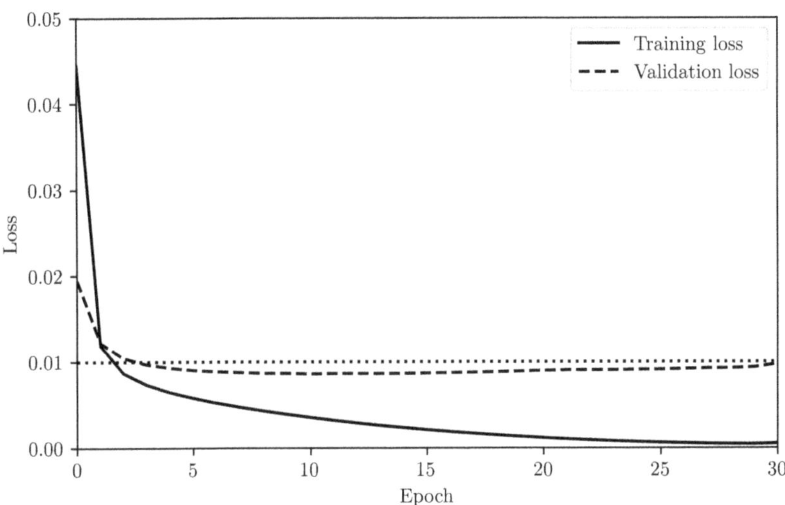

Fig. 6.2 Example for a model with a high capacity and no regularization that is overfitting

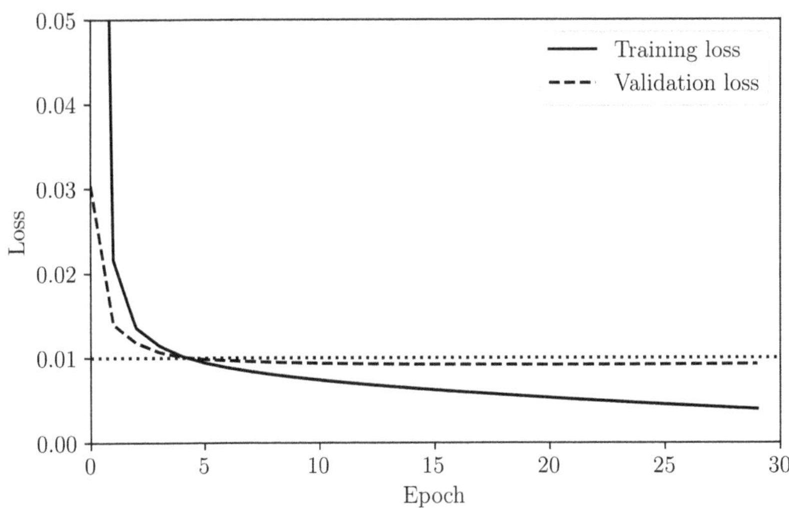

Fig. 6.3 Example for a model with a high capacity but regularization using dropout

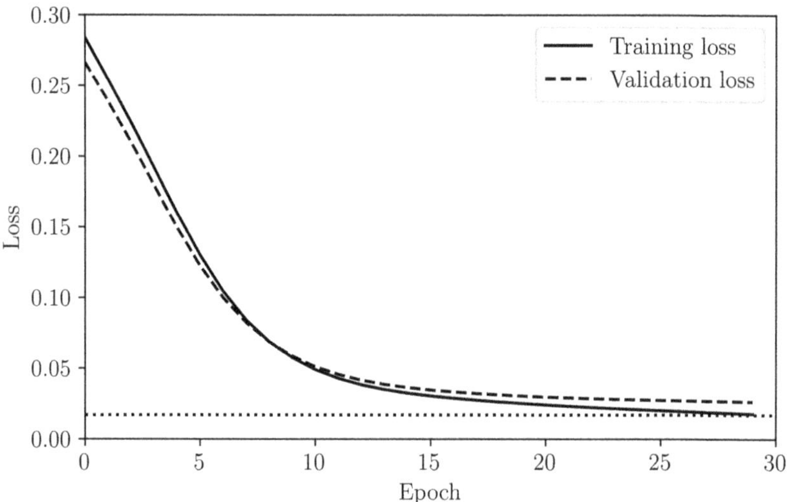

Fig. 6.4 Example for a model with a low capacity that is underfitting

curve is 0.009. This suggests that we could have improved the model's performance by stopping the training early.

Another regularization technique we introduce in Sect. 4.1 is known as dropout. To see the effect, we trained the same model as before, but added a dropout, with a dropout rate of 0.8. The results are depicted in Fig. 6.3. Adding dropout helped narrow the gap between the training and validation errors. Ultimately, the final validation error in this case was 0.08. By incorporating dropout, we achieved a model with enhanced generalization capabilities, thereby improving its ability to generalize beyond the training data.

Another concept we introduced is the concept of underfitting. We should expect the model to underfit if we use a model with a low representational capacity. Let us train a one-layer LSTM with a hidden size of 16, thus a model with a low capacity. The results are visualized in Fig. 6.4. Again we can see that the model improves every epoch. Also, we can observe that the gap between our training and validation errors is smaller than before. But we can also see that both our training and validation errors never get below 0.017. In Sect. 7.1, we will see that this is the MSE when guessing the average of our data. So this model's capacity is high enough to learn to predict the average value but too low to capture any underlying logic of emotions in music.

Hyperparameter Optimization

In order to further improve the performance of our model, we need to optimize the choice of our hyperparameters. This process is called **hyperparameter optimization**.

Since hyperparameters that control our model capacity have a U-shaped performance curve (see Sect. 4.1), the values must be chosen just right. As we have just seen, values that are too high or too low yield insufficient results. Finding the right parameters is challenging due to several reasons:

- **Computational expense**: Training and evaluating models with different hyperparameter settings is computationally expensive and time-consuming. Running a three-digit number of experiments with LSTM models, for example, can easily result in multiple days of computation.
- **Combinatorial Search Space**: Hyperparameters affect each other, resulting in many possible combinations to explore [19].
- **Lack of Intuition**: The impact of hyperparameters on model performance is often non-linear and non-intuitive. Small changes in hyperparameters can lead to significant variations in model performance, making it challenging to identify optimal settings [19].

Given a good starting point, such as one determined by others having worked on the same type of application and architecture (Table 6.2), one can try to manually change hyperparameters to improve the performance of the model. The first attempts at hyperparameter optimization in this thesis were also made by training the model, manually adapting the parameters, and repeating this pattern. While this can help to get a feeling for how the model behaves in different settings, it can be pretty laborious and might not always be the best way of tuning the model. Mainly due to the above-mentioned computational expense and the size of the search space, the author found this approach not to be expedient. One training run for the LSTM-model, for example, can take up to 30 minutes, depending on the configuration. As a matter of fact, tuning hyperparameters manually *can* work well, "when the user has months or years of experience in exploring hyperparameter values for neural networks applied to similar tasks" [19, p. 432]. Since the author of this thesis does not have the necessary experience or domain expertise, this approach was not suitable for this thesis.

Another approach is to apply algorithms to find optimal hyperparameters automatically. These **hyperparameter optimization algorithms** automatically try different hyperparameter combinations of a defined search space and find the best-performing combination. Each experiment with one combination will further be referred to as a **trial**. There are simple hyperparameter optimization algorithms such as grid- and random search or more complex model-based algorithms such as Bayesian optimization or genetic algorithms [19]. Additionally, these algorithms usually implement some kind of **scheduler** that terminates less promising trials after some epochs to use computational resources for more promising trials.

> **The Meta Role of Hyperparameter Optimization in Deep Learning**
> Hyperparameter optimization algorithms play a meta role in the realm of DL. While our DL-model *optimizes* its parameters (weights and biases) during training, these algorithms *optimize* the hyperparameters that define the learning process itself. Intriguingly, these hyperparameter optimization algorithms possess their own hyperparameters, which control the search space or search strategy.

Luckily, the research community has developed libraries with implementations of all of these algorithms. This thesis employs the Tune Model of the **Ray library** for hyperparameter optimization [36]. Ray enables parallel execution of several trials and integrates with different search algorithms. In this thesis, we use **Optuna Search** [2], which uses a tree-structured parzen estimator for choosing new combinations for each trial. The Asynchronous Successive Halving Algorithm (ASHA) is employed as a scheduler, enabling us to stop poorly performing trials early on. We set the grace period of ASHA to 10, meaning that each trial is given 10 epochs before being terminated. With this, we hope to give trials that might perform well after many epochs but learn slowly a chance while still efficiently using the computational resources.

Using this library makes it easy to implement these hyperparameter optimization algorithms. Luckily, we have all our training code wrapped in the `train_model` function, which receives a `config` containing the hyperparameters for a trial and reports the validation and test loss of each epoch back to the hyperparameter optimization algorithm. The hyperparameter optimization for the LSTM-model then can be implemented as:

```
1  param_space = {
2      "bi": tune.choice([True, False]),
3      "hidden_size": tune.choice([2**i for i in range(2,9)
       ]),
4      "lr": tune.loguniform(1e-6, 1e-3),
5      "batch_size": tune.choice([2**i for i in range(1,7)])
       ,
6      "num_layers": tune.choice([1, 2, 3, 4]),
7      "dropout_rate": tune.uniform(0, 1),
8      "epochs": 50,
9  }
10 scheduler = AsyncHyperBandScheduler(grace_period=10,
       max_t=50)
11 tuner = tune.Tuner(
12     train_lstm,
13     param_space=param_space,
14     tune_config=tune.TuneConfig(
15         metric="val_loss",
16         mode="min",
17         search_alg=OptunaSearch(),
18         scheduler=scheduler,
19         num_samples=200
20     ))
21
22 results = tuner.fit()
```

Code example 6.7 Using Ray to implement automatic hyperparameter optimization

With `param_space`, we define our hyperparameter search space; it consists of various parameters, some of which are presented as a list of choices, while others are represented as uniform or log uniform intervals. In line 11, we initialize the tuner by passing it the `train_lstm` function, which we aim to optimize, and the `param_space` we wish to search within. Additionally, we define our search algorithm and scheduler and state that we want to minimize the `val_loss` and limit the number of trials to 200.

When calling `tune.run()`, we start the search, which takes around one day to complete for the LSTM-model. Due to the lower complexity of the MERT-based model, the training and hyperparameter optimization for this model is much faster. At the end of the process, we receive a table with the results and configuration of all our trials. We do this procedure for both our models and use the configuration of the trial that performs best on the validation set. The hyperparameters that are used for model assessment and experiments in Chap. 7 are shown in Table 6.3 and Table 6.4.

Table 6.3 Final model configuration chosen after hyperparameter optimization for LSTM-model

Category	Value
Bidirectional	True
Number of Layers	3
Hidden Size	256
Batch Size	16
Learning Rate	$1.42 \cdot 10^{-4}$
Dropout Rate	0.35
Early Stop Threshold	1

Table 6.4 Final model configuration chosen after hyperparameter optimization for MERT model

Category	Value
Hidden Size	32
Batch Size	4
Learning Rate	$7.01 \cdot 10^{-3}$
Dropout Rate	0.52
Early Stop Threshold	2

In the previous Chapter, we developed two DL-models for the MER task—namely the LSTM model and the MERT model. The scope of this chapter is to assess their performance and to conduct some experiments regarding personalized MER.

After introducing some additional metrics, that will be used in this chapter, Sect. 7.1 will introduce a baseline and compare the performance of our two models on our test set. Section 7.2 uses the best-performing model architecture and conducts experiments on the personalization of MER. Section 7.3 summarizes the key findings of this Chapter and critically discusses them.

Metrics used in this chapter

In model development, we have used the MSE as a loss function and utilized this metric to evaluate the performance of various model configurations on our validation set. However, this chapter will introduce additional metrics to assess our model's performance.

While the MSE serves as a computationally efficient loss function, the Root Mean Square Error (RMSE) offers the advantage of being in the same *unit* as the output of our model. In our case, the *unit* is a normalized valence and arousal rating of a song, as explained in Sect. 6.1. Consequently, RMSE measures the model's average prediction error. Furthermore, RMSE is a metric that other researchers commonly use in MER, which can make results more comparable.

Nevertheless, comparability across datasets has limitations due to variations in dataset characteristics, including differences in variance. When making predictions on a dataset with a higher variance, the errors may be larger, potentially yielding a higher RMSE. We will utilize the R^2-**Score**, a metric that considers such factors to account for this variability. It is defined as

Y. Venohr, *Deep Learning in Personalized Music Emotion Recognition*, BestMasters, https://doi.org/10.1007/978-3-658-46997-9_7

$$R^2 = 1 - \frac{SS_{\text{err}}}{SS_{\text{tot}}}.$$

SS_{err} is the sum of squared errors defined as

$$SS_{\text{err}} = \sum_j (y_j - \hat{y}_j)^2$$

where y_j is the emotion value for song d_j and $\hat{y} = f(d_j)$ is the predicted value of our model f. SS_{tot} is the total sum of squares defined as

$$SS_{\text{tot}} = \sum_j (y_j - \bar{y})^2,$$

where $\bar{y}$ is the average emotion value over all songs. In less mathematical terms, R^2 can be seen as *the error of our model, compared to the variance of our data*. If we had a perfect model, SS_{err} equals 0 and R^2 results in 1. If we were predicting the average of our test set, the fraction becomes 1 and R^2 results in 0. Theoretically R^2 can also be negative, since SS_{err} can get arbitrarily big.

However, it is essential to note that these metrics are still not directly comparable across different datasets or subsets of a dataset. Each dataset may have its own unique characteristics and data distributions. Also, we can imagine that there are songs that are easier to predict than other songs. So there can be such a thing as an easy-to-predict dataset or subset of a dataset.

In addition to using these metrics for the vector $y \in \mathbb{R}^2$ representing our dimensional emotion value, we will use these metrics on the valence and arousal scalar values individually.

It is important to emphasize that, whereas low values of RMSE and MSE indicate good performance, a high value of the R^2-score indicates good performance.

7.1   General Model

Looking back at Sect. 6.2, we differentiated between model development and model assessment. While that chapter had the scope of developing and optimizing our two models, the scope of this section is the assessment of them.

In the previous chapter, we trained the models on the train split D_{train} and evaluated their ability to generalize on the validation split D_{val}. However, the test split has remained untouched and has yet to be utilized for model assessment. In order to thoroughly evaluate our final models, we will employ this test split D_{test}. Since the fine-tuning process has concluded, the validation split is no longer required, allowing us to repurpose these additional data points for training our model. Thus we can train on our models on $D_{\text{train/val}} = D_{\text{train}} \cup D_{\text{val}}$.

Defining a Baseline

When developing or assessing any machine learning model, it is crucial to determine whether the model can learn meaningful patterns from the data. Since it is hard to evaluate a MSE without any point of reference, it is common practice to define one. This point of reference for model evaluation is usually called the **baseline**. The baseline is the initial benchmark against which the model's results can be measured.

For this thesis, we have chosen to set the average valence and arousal values of the test set as our baseline. The idea behind this choice is that it provides a simple yet informative benchmark that represents a reasonable prediction for the target values. Our baseline model will predict these average values for each song in the test set. The prediction of our baseline model $f_{\text{baseline}}(d_j) = \hat{y}$ will be

$$\hat{y} = \overline{y}_{\text{test}} \forall j \in J_{\text{test}},$$

where $\overline{y}_{\text{test}}$ is the average values of valence and arousal of our test set and J_{test} are the indices of the songs in our test set D_{test}.

Another approach could have been to use random values as a baseline. However, this would not be representative of the data distribution, since the data in our dataset is not uniformly distributed (see Fig. 2.6).

Furthermore, it is common practice to use the results of previous research on the same dataset as additional reference points. Since DEAM is mainly designed as a benchmark for MEVD, the author could not find any results for static MER on the full DEAM. Thus, this approach is not suitable for this thesis.

However, we can use the results of the best-performing teams in the static task of the 2013 MediaEval competition as an orientation. These are the results for a subset of 1000 songs of the DEAM, as presented by Soleymani et al. [50]. Again, even though they can serve as an orientation, they are not directly comparable to our results since they are evaluated on a different subset. The results can be seen in Table 7.1.

Table 7.1 Summary of results for the static task in the 2013 MediaEval challenge [50]. TUM: TU Munich, UoA: University of Aizu, UU: Utrecht University

Group	RMSE Valence	RMSE Arousal	R^2 Valence	R^2 Arousal
TUM	.11	.10	**.42**	.59
UoA	.12	.10	.35	**.63**
UU	.12	.10	.31	.59

Quantitative Results

The results of our general MER models, in comparison to the previously described baseline, is presented in Table 7.2.

Table 7.2 Summary of Results of different models for general MER

Model	RMSE Valence	RMSE Arousal	R^2 Valence	R^2 Arousal
Baseline	.122	.140	0	0
LSTM	.097	.097	.373	.518
MERT	**.088**	**.084**	**.485**	**.639**

First, let us look at the results of our baseline. 0.122 and 0.140 are exactly the standard deviation σ_{test} of our test set as we have seen in Sect. 6.1. This is not surprising when we consider the definition of the RMSE and the standard deviation. Due to our choice of the baseline as guessing the average, calculating the RMSE on our baseline equals calculating the standard deviation of our test set. Also, the results for R^2 are not surprising if we look at the given definition.

Looking at the results of our models, we can see that both models outperform our baseline in all metrics, which means they are both better than guessing the average emotion of a song. This indicates that both models have captured some meaningful pattern or relationship of emotions in music.

Looking at the metrics, we can make another observation. In both models, the RMSE for valence and arousal is similar (e.g. $0.088 \approx 0.084$). Still, there is a bigger difference in the R^2 score between them (e.g. $0.485 \not\approx 0.639$). This could be explained due to the different standard deviations of valence and arousal in our test set. Since the valence dimension has a smaller standard deviation (and thus variance) than the arousal dimension, the SS_{tot} term becomes smaller, leading to a smaller R^2 score.

Comparing the two models, we can see that the MERT model outperforms the LSTM model in all metrics. The average error (RMSE) of this model is 0.088 for the valence of a song and 0.084 for the arousal of a song. With respect to the original nine-point rating, this is equivalent to an error of 0.79 and 0.76, which seems quite low. On the other hand, evaluating a RMSE highly depends on the test sets distribution. Better insight into our model's performance can be gathered by looking at the R^2 score. The MERT model achieved an R^2 score of 0.485 for valence and 0.639 meaning that our model can explain 49% and 64% of the variation of our data.

Figure 7.1 visualizes the distribution of the error for valence and arousal as a histogram with a kernel density estimation [1]. The strong resemblance of the kernel density estimation to a normal distribution indicates that our model has no bias error.

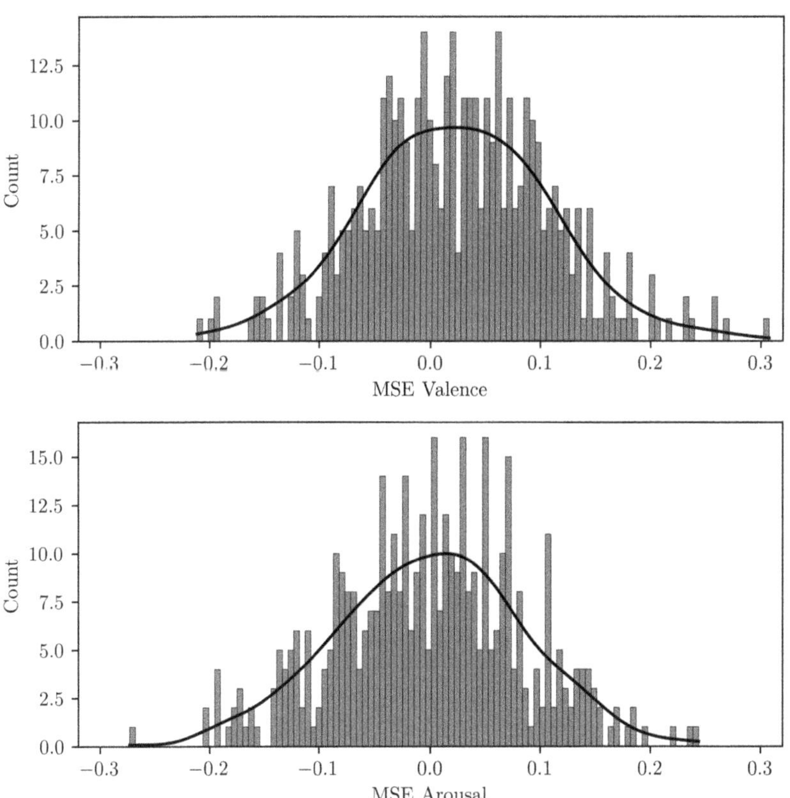

Fig. 7.1 Distribution of valence and arousal error of MERT model on D_{test}. Additionally to the histogram, a kernel density estimation [1] is plotted

When comparing to the 2013 MediaEval results, we can observe that the MERT model performs similar or better than the best results. At the same time, we have to acknowledge that the results are not comparable since they are training and testing on a subset of DEAM and thus also had less data to learn from.

For the experiments that are conducted in the rest of this chapter, we will utilize the MERT model due to its superior performance.

Qualitative Results

In order to supplement the evaluation using numerical metrics and to gain deeper insights into the MERT model's performance, the author conducted a qualitative assessment. This was done by listening to the songs of the test set and evaluating the predicted and actual valence and arousal of the songs. It is important to note that this assessment is based on personal judgment and impressions rather than being systematic and rigorous. By doing this, we still hope to gain a better *feeling* for the model's performance since it is easy to get lost in optimizing abstract metrics such as the R^2-score. The values given in braces are the ids of the referenced pieces as defined in DEAM.

A notable observation mirroring the quantitative findings was the low variance of the data, which was discussed in Sect. 2.3. Many songs exhibited moderate valence and arousal levels, particularly within the pop (436) and hip-hop (407) genres. This made it hard to do qualitative observations. The analysis turned to songs with strong emotional characteristics to overcome this limitation.

A set of pieces that yielded noticeable emotional cues were those featuring acoustic guitars. While several slow and melancholic guitar pieces in minor keys were observed (50, 68, 95), the model consistently recognized their low valence. While one could assume the model learned a simple *guitar equals sadness* notion, song 96 proved this assumption wrong. Song 96, a playful and swinging acoustic guitar song, was successfully detected with high valence and arousal.

On the other hand, there were some fast-paced, happy rock songs (39, 345, 72) where the model successfully detected high valence and arousal. This shows the model's ability to capture their energetic and upbeat nature. This held for various genres, including a drum and bass mix (395) and a 70s disco beat (466).

Piano pieces also showed some subtleties that the model was able to capture. For instance, songs 146 and 153 both featured slow and calming piano melodies that the model accurately described with a low arousal. Yet the model accurately distinguished 153 as having a more hopeful tone, resulting in a higher valence. Song 254 served as an example where the model successfully detected high valence and arousal in a piano piece.

Some experimental songs (44) posed a challenge for the model in predicting valence. However, the author also found it difficult to ascertain whether these songs conveyed happiness or sadness, highlighting the subjective nature of interpreting such songs.

In quieter compositions, such as song 69, the model accurately detected low arousal. Notably, song 281, an old and quiet recording of an upbeat band, showcased the model's ability to recognize the underlying high arousal despite the quiet nature of the recording.

Effect of Training Data Amount

A fundamental principle in machine learning is the belief that the model improves with more training data. Let us try to see if the performance decreases if we use less training data. For this, we trained models with different amount of training data and tested them on the same test set. Let us define

$$D_{\text{train/val},N} := \{(d_j, y_j) : \text{ for the first } N \text{ indices } j \text{ in } J_{\text{train/val}}\}$$

as a subset of $D_{\text{train/val}}$ by just using the first N indices of $J_{\text{train/val}}$. Then we train models f_N for some different values of $N = 50, 100, ..., 1250$ and calculate the MSE loss of f_N on the test set D_{test}. The results are shown in Fig. 7.2. We can see N on the x-axis and and MSE of f_N on the y-axis.

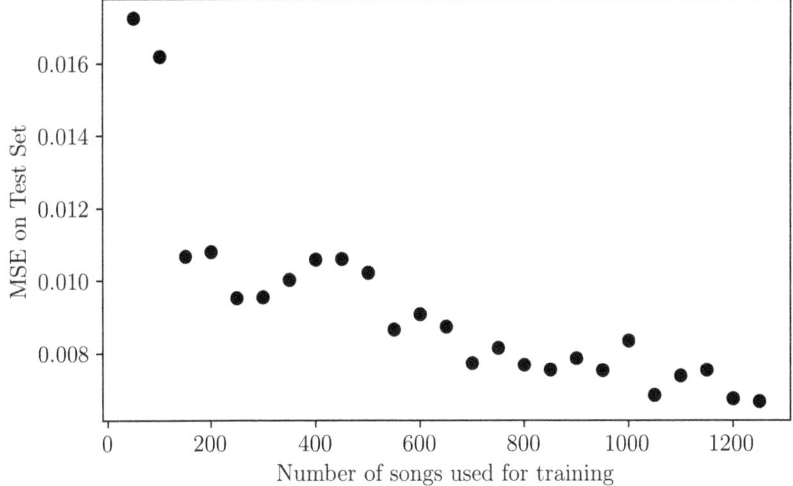

Fig. 7.2 MSE loss on the test set for models that have been trained on a different number of songs

At first glance at Fig. 7.2, we can say that the performance of our model indeed decreases if we use less training data. There is a clear trend of the error decreasing when N is increased, and the best performance is reached using all available data points. Furthermore, we can see a strong performance increase when the sample size N is larger than 100. This could represent a minimum amount of examples for our model to learn anything meaningful about the relation of music and emotions.

Interestingly, we can observe that the decrease of MSE is not continuous. This could be explained by the inherent variability of the training data, meaning that some of the added 50 samples are a biased representation of the entire population, which could lead to a decrease in performance. Another explanation could be that there is something like *bad learning example*. Not all song examples contribute equally to the learning process, and some may even introduce noise or confusion. Possibly, there are song samples that are misleading for the model. Similarly to human learning, where the choice of learning examples can sometimes lead to confusion, machine learning models can also be influenced by misleading or inappropriate data. Thus, it is essential in machine learning to consider the quality and relevance of the data used for training.

7.2   Personalization

In the previous section, we developed a model architecture and fine-tuned it to detect emotions in music. We trained the model on a ground truth defined by the averaged emotion annotations of all annotations for one song. In other words, we built a general MER model. Additionally, in Sect. 2.2, we discussed the limitations of general MER models. Even a perfect model would merely predict the average emotion and not the perception of an individual user. Especially when considering the low inter-rater agreement on valence and arousal, the author believes that the most exciting direction to examine further is the personalization of MER models.

Learning Personal Nuances: Experiment
Firstly, we will experiment to see whether our model can learn the personal nuances of emotions in music. We will use the personal annotations provided in the DEAM and build a personalized MER model. In the following, we will examine the ten subjects that annotated the most songs (see Sect. 2.3). Table 7.3 shows the worker ids of the ten subjects and their short ids that will be used in the following.

To answer whether our model can learn the personal nuances of emotions in music, we will train three models for each subject s_i. Let us remember that we have our dataset D of all songs d_j with indices $j \in J$.

Table 7.3 Annotator Ids and short ids of subjects that will be used for experiments

Id	Annotator Id	N Songs
1	eca1130f44cd2e17645e40a0fa2ef59b	1005
2	a30d244141cb2f51e0803e79bc4bd147	985
3	607f6e34a0b5923333f6b16d3a59cc98	955
4	de2b2c35312ac2f0a8510743742c0219	937
5	2afd218c3aecb6828d2be327f8b9c46f	760
6	00de940f0b5cfc82cca4791199e3bfb3	751
7	78b5e9744073532cc376976b5fc6b2fc	718
8	b09a5957e5d5e47e556d203529a0ae6d	708
9	ed7ed76453bd846859f5e6b9149df276	675
10	ff9c1993d2a21f2117c30d8e295dd4ac	661

Let J_i be the indices of all songs d_j, that have been annotated by subject s_i. Now let D_i be the subset of D consisting of the songs with indices J_i. The number of samples in D_i can be seen in Table 7.3.

This subset D_i is then randomly split into a test set $D_{i,\text{test}}$ (25%) and train set $D_{i,\text{train}}$ (75%) with their canonical set of indices $J_{i,\text{test}}$ and $J_{i,\text{train}}$.

For each of the ten subjects s_i we have three models:

- **Personalized model:** Trained with $\{(y_{ji}, d_j) : \forall j \in J_{i,\text{train}}\}$, where y_{ji} is the emotion annotation of subject s_i for song d_j.
- **General model:** Trained with $\{(\overline{y}_j^{(\backslash i)}, d_j) : \forall j \in J \setminus J_{i,\text{test}}\}$, where $\overline{y}_j^{(\backslash i)}$ is the average emotion annotation of all subjects, except for the annotation of subject s_i for song d_j.
- **Perfect general model:** Always predicts $\hat{y}(d_j) = \overline{y}_j$.

All models are tested on $\{(y_{ji}, d_j) : \forall j \in J_{i,\text{test}}\}$.

Let us clarify our three scenarios. The purely personalized model has the advantage of being trained on personalized annotations while having the disadvantage of not being trained on too many samples. For example, Subject 1 is only trained on $1005 \cdot 0.75 \approx 754$ samples. Then we have a general model with small limitations. Firstly, it is not trained on the songs that are in $D_{i,\text{test}}$. In a real-world context, this is the scenario that we want to make a personal emotional prediction for an unknown song. For Subject 1, this model is still trained on $1744 - (1005 \cdot 0.25) \approx 1493$ samples. The second limitation is that we are considering the scenario that the model does not know the annotations of subject s_i. This means that we calculate the average of all annotations except for the annotation of s_i. In Sect. 2.3, we mentioned that every song has been annotated by at least ten subjects (not to be confused with the ten subjects we are using for this experiment). Concluding we are building the average of at least nine other subjects. Lastly, we have the perfect general model. In this case, we are imagining an almost omniscient model that can perfectly predict the average emotion of a song.

Learning Personal Nuances: Results

The results of this experiment are summarized in Table 7.4. The first observation is that the personalized model outperforms the general model for all ten subjects. This is impressive because the general models have been trained on at least double the number of songs than the personalized model. As we have seen in Sect. 7.1, the number of songs the model is trained on strongly affects the model's performance. We conclude that the model can learn personal nuances of emotions in music with enough training data.

Table 7.4 Summary of Results for personalized models

Subject	Model	RMSE	R^2 Valence	R^2 Arousal
1	**personalized**	**.097**	.374	.613
	general	.105	.304	.529
	perfect general	.101	.231	.625
2	personalized	.174	.138	.167
	general	.175	.181	.111
	perfect general	**.170**	.324	.036
3	**personalized**	**.125**	−.094	.271
	general	.147	−1.259	.25
	perfect general	.133	−.808	.387
4	**personalized**	**.213**	.589	.616
	general	.286	.275	.291
	perfect general	.265	.278	.479
5	personalized	.118	.653	.197
	general	.135	.475	.091
	perfect general	**.112**	.719	.195
6	**personalized**	**.144**	.055	.263
	general	.209	−.269	−1.526
	perfect general	.179	.01	−.725

(Continued)

Table 7.4 (Continued)

Subject	Model	RMSE	R^2 Valence	R^2 Arousal
7	personalized	.182	.032	.229
	general	.205	.008	−.267
	perfect general	**.180**	.306	−.081
8	personalized	.168	.13	.553
	general	.187	−.062	.44
	perfect general	**.152**	.32	.624
9	personalized	.178	−.015	.384
	general	.182	−.223	.444
	perfect general	**.157**	.09	.59
10	**personalized**	**.100**	.26	.553
	general	.149	−.252	−.294
	perfect general	.138	−.257	.022

The second observation is that the personalized model even outperforms the perfect general model for half of the cases. Even with the limited training data, the personalized model can capture personal nuances and make better predictions than the omniscient general model. It is surprising how bad the perfect general model performs for some of these individuals. This can be explained by the low inter-rater agreement we introduced in Sect. 2.3. Even a perfect general model only predicts the average emotion of a song. And if individuals strongly disagree with each other, the prediction of a general model is bad for the individual.

Another observation is that the performance of the general model strongly differs between subjects. A possible explanation could be that some subjects disagree stronger with the average opinion than others. For example, for subject six in the arousal dimension, the perfect general model got an R^2-score of -0.725 while it got 0.613 for subject one. It seems like subject six has a stronger disagreement with the arousal ratings of the average annotator than subject one. It would be interesting to see whether this is because subject six is more enthusiastic in using the extreme values on the rating scale and the personalized model learned this or if there might be specific cues in music that affect the arousal in subject six when listening to music.

Model Averaging Approach

The perfect general model outperforms the personalized model for the other half of the subjects. This means that the training amount was insufficient for the personalized model to learn to predict as well as the perfect general model. We assumed, that these individuals may have an emotional reaction more in line with the average emotional reaction.

In this case, we can still outperform the perfect general model by using a simple personalization technique called the **model averaging approach** we introduced in Sect. 2.2. This technique combines predictions of a general model trained on annotations from a broad user base and a personal model trained on annotations by the target user. For this we introduce a weighting parameter $\omega \in [0, 1]$, that determines the **degree of personalization** and define our new prediction as:

$$\hat{y}_{combined} = (1 - \omega) \cdot \hat{y}_{general} + \omega \cdot \hat{y}_{personalized}$$

The results for Subject 2 can be seen in Table 7.5. It is fascinating that the truth actually lies between our general model's predictions and our personalised model's predictions. With a degree of personalization of .5, we can outperform the perfect general model by combining the large annotation base of the general model and the captured personal nuances of the personalized model.

Table 7.5 Varying the degree of personalization ω for Subject 2

ω	RMSE	R^2 Valence	R^2 Arousal
w=.0	.175	.181	.111
w=.1	.173	.196	.132
w=.2	.171	.208	.15
w=.3	.170	.215	.164
w=.4	.170	.217	.175
w=.5	.169	.215	.182
w=.6	.170	.208	.186
w=.7	.170	.197	.187
w=.8	.171	.182	.183
w=.9	.173	.162	.177
w=1	.174	.138	.167

We have just seen that our model can identify personal nuances, and we showed that a personalized MER model outperforms a general MER model on unseen songs all the time. If we already know the averaged emotion of a song (or have a perfect general MER model), it depends on the individual whether personalized MER is better than a general MER. Nevertheless, even in this case, we have shown that we can achieve a better prediction using a combination of the general and personalized models.

Reducing User Burden using Residual Modeling

The biggest challenge in personalized MER is the availability of personalized annotations. Previously, we were using around 500 annotations to train the personalized model. While using all these annotations and using the model averaging approach, we were able to outperformed even a perfect general model, this approach is not operable in real-world scenarios since it will be hard to find a user that is willing to give that many annotations.

After repeating the experiment we have done previously with only 100 personal annotations, we observed that the personalized model outperforms the general model only for five out of ten subjects (more results in Repository). This shows that when using only 100 personal annotations, our model cannot learn enough about the individual to outperform a general model. So building a personalized model, as we have done previously, is not a suitable option in practice.

For the last experiment, we utilize another personalization technique called **residual modelling**, which we introduced in Sect. 2.2. In addition to a general model trained on annotations from a broad user base, we will train another model to predict the **perception residual** between the general perception and the individual one. Given a music piece d_j and an individual s_i, we define the perception residual $y_{ij}^{\Delta} := y_{ij} - y_j$ as the difference between the perceived emotion y_{ij} of subject s_i and the averaged perceived emotion y_j of this musical piece. Then for each subject s_i we train the model with

$$\{(y_{ij}^{\Delta}, d_j) : \forall j \in J_{i,\text{train}}\}.$$

By doing this, we hope to get a model that can predict the perception residual $\hat{y}_{ij}^{\Delta}$ of an individual s_i for a given song d_j. Since we expect this approach to work even with less training data, we train another model similar to the one we just described but limit the training songs to 100. The final prediction $\hat{y}_{\text{residual}}$ of our residual model will then be defined as

$$\hat{y}_{\text{residual}} = \hat{y}_{\text{general}} + \hat{y}^{\Delta}.$$

Based on the assumption that similar aspects of a song influence the perception residual, we utilize the same MERT model architecture as we used to predict the perceived general emotion in our previous experiments.

However, we observe a rather unexpected outcome upon analyzing the results presented in Table 7.6. Instead of improving the predictions, adding the predicted residual led to worse results. Surprisingly, the model trained on a limited dataset of 100 songs even outperformed the residual model trained on all available songs. These results are not aligned with our expectations. Firstly, we expected that adding the predicted residual would enhance the predictions of the general model. Secondly, we expect the residual model to improve with more training data.

Table 7.6 Results for residual modelling approach for subject 1

Subject	Model	RMSE	R^2 Valence	R^2 Arousal
1	**personalized**	**.097**	.374	.613
	general	.105	.304	.529
	residual modelling	.123	.154	.272
	residual modelling 100	.116	.208	.369
	perfect general	.101	.231	.625

Several factors may contribute to these unexpected results. One possibility is that the residual itself contains a significant amount of noise. Altering certain hyperparameters could have addressed this issue, but unfortunately, we lacked the time to analyze and experiment with different settings thoroughly.

7.3   Discussion

In this section, we summarize the key findings of this chapter and critically discuss some aspects of it.

In the scope of this thesis, two MER models were developed and evaluated. One with a LSTM-architecture, that was using handcrafted audio features and one using embeddings from the music understanding model MERT and CNN + FC layers for regression. Section 7.1 evaluated these two model architectures. Both models outperformed our baseline of random guessing. Thus we conclude that they were able to capture meaningful patterns or underlying logic in the data they were trained on. To gain further confidence in the model's performance, comparing it to a more elaborate baseline, such as support vector regression, could have been valuable. For both models, we found that predictions on the arousal dimension achieved better results than predictions on the valence dimension. This aligns with previous research findings that arousal is easier to model than valence.

On comparison of the two models, we observed that the MERT model was better at detecting emotions in music than the LSTM model. It achieved a smaller generalization error in both the valence and arousal dimensions. Since the two models are based on entirely different architectures, it is hard to pin down a specific reason for the differences in performance. It is important to note that the LSTM model was trained on a time series of features and thus was explicitly using the time dimension. On the other hand, the regression head of the MERT model uses the averaged vector representation of the songs, thus not explicitly modelling time dependencies. Since MERT was trained to generate its representations with a context of 5 seconds, the information about time is implicitly regarded in this model. Combining the two approaches could be interesting. One could imagine to extract embeddings for smaller time instances then using these to train an LSTM model.

Furthermore, the design and implementation of the LSTM model also involved much more decisions compared to the MERT model. Even though there are research papers describing the implementation of their models briefly, the author was not able to find any publicly available implementations of it. While writing this thesis, the author gained some understanding of how this architecture works in theory, there are still many details to consider when implementing a model like this. Thus, the bad performance of the LSTM model could be caused by misconceptions in implementing the theoretical concepts.

However, using the embeddings of the music understanding model MERT, the author was able to train a model that could capture the valence and arousal of musical pieces. It could explain 49% and 64% of the variation of valence and arousal. This is comparable to other static MER research using data from DEAM. When analyzing the predictions of the model in a qualitative way, the author found many instances where the model was able to capture fine nuances in the perceived valence and arousal of musical pieces. It is important to note that while an informal qualitative analysis can provide valuable insights and supplement the assessment using numerical metrics, it may lack the objectivity and replicability of a more systematic approach.

Even though we were able to underline the good results qualitatively, it is still important to consider data distribution and possible biases in the test set when interpreting results. For this thesis, we used a fixed test split of 25% that stayed locked away in model development. However, to gain more confidence in the model's performance, it would have been interesting to test its performance on different splits using techniques such as k-fold cross-validation.

In Sect. 7.1, we could show that our model performance increases with the amount of data it is trained on. This aligned with our expectations. At the same time, we could observe that this increase was not continuous. Likely this can be explained with training examples that are rather confusing than helpful for the model to learn about the underlying relation between music and emotions. These observations raise interesting questions for potential further research. Firstly, how far can we increase the performance of a general MER model if we collect more data. Secondly, how can we identify *good* and *bad* learning examples. It would be interesting to systematically quantify each song's learning effect while always testing on an unbiased test set.

Due to the subjective nature of emotions in music, the performance of a general MER system is inherently limited. Even a perfect general model only predicts the average emotion of a song. And if individuals strongly disagree with each other, the prediction of a general model is bad for the individual. This effect was shown in Sect. 7.2. We were able to show that, given enough annotations, the personal model always outperforms a general model. More interestingly, we were able to show that it is even possible to outperform a perfect general model utilizing personalisation techniques. The best performance was achieved when making use of the large annotation base of the general model and the captured personal nuances of a personalized model and combining them using the model averaging approach.

Even though this comparison highlights the superiority of personalized models over general models, the comparison is of a rather theoretical nature. On the one hand, collecting 500 annotations for a personalized MER system is impractical due to the high user burden involved with this process. On the other hand, we will not be able to develop a truly perfect general MER or manually annotate *all* songs in the world.

One aspect that has yet to be critically discussed and quantified is that different annotators may have similar emotional responses but use the rating scale differently. For example, one person might use values between four and six, while another is more enthusiastic about using very high and low values on the rating scale. For example, a subject that annotates the same direction of emotions as the average but in different extends on the scale. In this case, the personalised model's performance would exceed the general model's performance when looking at metrics as R^2-score, while qualitatively not being better. It would be interesting to normalize each individual's answers beforehand to ensure a fair comparison between the general and personalized models.

In the last experiment, we trained a model using the residual modelling approach. The results show that incorporating the predicted residual into the model did not lead to the anticipated improvements. This does not align with the findings of Yang [53], that a residual model with just 10 personalized annotations can improve results significantly. Here, further investigation is required to understand the reasons behind these outcomes fully and to explore alternative approaches that could address the challenges posed by the noisy nature of the residual. Additionally, it is worth noting that the residual values were relatively small in magnitude. Applying scaling techniques to the residual could have led to more favourable results. However, we could not explore this in detail due to time constraints.

8

As we conclude this thesis, let us revisit the research questions we formulated in the introduction:

- **RQ 1:** How can we develop a mathematical model that can predict the emotional content of a musical piece?
- **RQ 2:** Can utilizing embeddings from the pre-trained music understanding model MERT enhance the prediction results?
- **RQ 3:** Is it possible for this model to learn individual nuances when trained on personalized annotations? If so, how can we minimize the user burden of collecting individual annotations?

Research Question 1

Scholars of many fields have researched how to build mathematical models to learn the relationship between music and emotions. Throughout this thesis, we explored different aspects of this research systematically.

For a mathematical model to learn the relationship between music and emotions, we need mathematical representations of the two. While introducing the foundations of MER in Chap. 2, we introduced Russel's model that describes the perceived emotion of a musical piece in two numerical dimensions—namely Valence and Arousal. In Chap. 3, we bridged the gap between the *sound of music* and its mathematical representation by introducing the time-pressure representation and time-frequency representations of it. Furthermore, we looked into handcrafted feature extraction as a method to extract emotionally relevant features from music.

This prework enabled us to regard MER as a regression problem. After introducing machine learning as a way of solving regression problems, we delved into the world of deep learning in Chap. 4. Firstly we focused on supervised deep learning and introduced neural networks and other architectures such as LSTM that can

Y. Venohr, *Deep Learning in Personalized Music Emotion Recognition*, BestMasters, https://doi.org/10.1007/978-3-658-46997-9_8

handle sequenced data. Then, we considered very recent advancements and introduced the concept of self-supervised pre-training as a way to generate music understanding models.

In Chap. 5, we consolidated these foundations and reviewed different approaches to the MER task. The traditional approach to MER involves handcrafted features, and machine learning algorithms like SVR. Recent advancements on the other hand have seen the adoption of deep learning architectures, such as LSTM, CNN and FC into their MER pipeline. Additionally, we explored how, even more recently, researchers began to utilize self-supervised pre-training on large unlabeled music corpora to build models that learn to create meaningful vector representations (embeddings) of a musical piece.

Based on this, two MER models were developed: one using handcrafted features and an LSTM architecture, and the other leveraging embeddings of the music understanding model MERT. Details of these models were described in detail in Chap. 6.

Research Question 2

In Chap. 7, these models were evaluated for their ability to predict the emotional content of music. While both models learned meaningful patterns, the experiments conducted in this thesis indicate that utilizing embeddings from the MERT model enhanced the prediction results compared to using handcrafted features with an LSTM architecture. Although the exact reasons for this improvement are not yet fully understood, the results obtained are comparable to those achieved by other authors on similar datasets.

With the MERT model, we were able to create a mathematical model capable of detecting a musical piece's valence and arousal. This highlights the potential of utilizing self-supervised learned representations for advancing the field of MER.

Research Question 3

In addition to the experiments with the general MER model in Chap. 7, this thesis used the individual annotations in the DEAM for creating personalized MER models. To our knowledge, this is the first time such an approach has been undertaken. We trained personalized models for the ten individuals who had annotated the most songs and employed a model-averaging approach to make personalized predictions.

Despite the inherent limitations in generalizing from training data, the personalized models could learn individual nuances and yield better results even compared to a perfect general MER. This highlights the superiority of personalized models.

Due to time constraints, we were not able achieve the desired results, concerning personalized MER with a limited amount of personal annotations.

Outlook

While this thesis has provided valuable insights into the topic of *deep learning in personalized music emotion recognition*, it has also revealed a range of questions that stayed unanswered and created the need for future investigations.

When observing the model's performance in relation to the amount of training data, two questions arose. To answer these, future research could explore the upper limits of model performance with increased training data and investigate methods to quantify the learning effect of each song systematically.

In order to supplement the assessment using numerical metrics, the author of this thesis conducted an informal qualitative review of the model's predictions. Since this approach lacked the objectivity and replicability of a more systematic approach, further research could profit from assessing existing models systematically in a qualitative way. In general, the author believes that the field of MER, driven by the adoption of deep learning techniques, has diverged from musicology. Moving forward, research could aim to establish stronger connections between work from experts in musicology and experts from the fields of machine learning. Additionally, while novel deep learning approaches have faced criticism for their *black-box* nature, future studies could also strive to achieve state-of-the-art results while maintaining model explainability and interpretability.

Additionally, the field of MER still faces challenges regarding publicly available high-quality datasets for research. In the process of collecting datasets for future MER research, it is essential to address potential biases. When collecting data, researchers should ensure that the data represents the cultural diversity of music listeners worldwide.

Due to the high subjectivity of emotions in music, the author of this thesis believes improving the performance of general MER models should not be the primary research direction. Instead, embracing subjectivity and focusing on personalized emotion recognition should be a priority. Consequently, research efforts could concentrate on finding solutions that enhance general models' performance even with a few personal annotations. Another aspect to consider in this research direction is ensuring a fair comparison between general and personalized models. Research could explore the possibility of different annotators having similar emotional responses but using the rating scale in different ways. Future research could explore techniques to normalize individual ratings to address potential differences in rating scale usage.

A promising way of enhancing the performance of general models, even with a small number of personal annotations, is residual modelling. Even though, due to time constraints, we could not achieve the desired results, further investigation could help to understand the reasons behind these outcomes fully and to explore alternative approaches that could address the challenges posed by the noisy nature of the residual.

Finally, it is important to underline that the models developed to this day capture only a simplified representation of emotions and focus solely on perceived emotions. Thus, it is crucial to recognize that mathematics, indeed, has little to say about the indescribable world of vivid memories and complex emotions that music can evoke in us. May we acknowledge that some aspects of human experience resist quantification and leave space for mystery and fascination.

Bibliography

1. A. Bowman and A. Azzalini, "Applied smoothing techniques for data analysis: The kernel approach with s-plus illustrations," *Journal of the American Statistical Association*, 1999.
2. T. Akiba, S. Sano, T. Yanase, T. Ohta, and M. Koyama, "Optuna," in *Proceedings of the 25th ACM SIGKDD International Conference on Knowledge Discovery & Data Mining*, A. Teredesai, Ed., ser. ACM Digital Library, New York, NY, United States: Association for Computing Machinery, 2019, pp. 2623–2631, ISBN: 9781450362016. https://doi.org/10.1145/3292500.3330701.
3. A. Aljanaki, Y.-H. Yang, and M. Soleymani, "Developing a benchmark for emotional analysis of music," *PloS one*, vol. 12, no. 3, e0173392, 2017. https://doi.org/10.1371/journal.pone.0173392.
4. L.-L. Balkwill and W. F. Thompson, "A cross-cultural investigation of the perception of emotion in music: Psychophysical and cultural cues," *Music Perception*, vol. 17, no. 1, pp. 43–64, 1999, ISSN: 0730-7829. https://doi.org/10.2307/40285811.
5. D. J. Benson, *Music: A mathematical offering*, 6th print. Cambridge: Cambridge Univ. Press, 2013, ISBN: 9780521853873.
6. M. M. Bradley and P. J. Lang, "Measuring emotion: The selfassessment manikin and the semantic differential," *Journal of Behavior Therapy and Experimental Psychiatry*, vol. 25, no. 1, pp. 49–59, 1994, ISSN: 0005-7916. https://doi.org/10.1016/0005-7916(94)90063-9.
7. Z. Che, S. Purushotham, K. Cho, D. Sontag, and Y. Liu, "Recurrent neural networks for multivariate time series with missing values," *Scientific Reports*, vol. 8, no. 1, p. 6085, 2018, ISSN: 2045-2322. https://doi.org/10.1038/s41598-018-24271-9.
8. S. Chowdhury, A. Vall, V. Haunschmid, and G. Widmer, *Towards explainable music emotion recognition: The route via mid-level features*.
9. E. Coutinho, G. Trigeorgis, S. Zafeiriou, *et al.*, "Automatically estimating emotion in music with deep long-short term memory recurrent neural networks," *CEUR Workshop Proceedings*, vol. 1436, 2015.
10. E. Coutinho, F. Weninger, B. Schuller, and K. R. Scherer, "The munich lstm-rnn approach to the mediaeval 2014 "emotion in music" task," *CEUR Workshop Proceedings*, vol. 1263, 2014.
11. A. Crame, H. Wu, J. Salamon, and J. Bello, "Look, listen, and learn more: Design choices for deep audio embeddings," in *ICASSP 2019—2019 IEEE International Conference on*

Acoustics, Speech and Signal Processing (ICASSP), 2019, pp. 3852–3856, ISBN: 2379-190X. https://doi.org/10.1109/ICASSP.2019.8682475.

12. Y. Dong, X. Yang, X. Zhao, and J. Li, "Bidirectional convolutional recurrent sparse network (bcrsn): An efficient model for music emotion recognition," *IEEE Transactions on Multimedia*, vol. 21, no. 12, pp. 3150–3163, 2019, ISSN: 1520-9210. https://doi.org/10.1109/TMM.2019.2918739.

13. T. Eerola and J. K. Vuoskoski, "A review of music and emotion studies: Approaches, emotion models, and stimuli," *Music Perception*, vol. 30, no. 3, pp. 307–340, 2013, ISSN: 0730-7829. https://doi.org/10.1525/mp.2012.30.3.307.

14. P. Ekman, "An argument for basic emotions," *Cognition and Emotion*, vol. 6, no. 3–4, pp. 169–200, 1992, ISSN: 0269-9931. https://doi.org/10.1080/02699939208411068.

15. F. Eyben, F. Weninger, and B. Schuller, "The tum approach to the mediaeval music emotion task using generic affective audio features," *Working Notes Proceedings of the MediaEval 2013 Workshop*, vol. Barcelona, Spain, October 18–19, 2013.

16. F. Eyben, F. Weninger, F. Gross, and B. Schuller, "Recent developments in opensmile, the munich open-source multimedia feature extractor," in *Proceedings of the 21st ACM international conference on Multimedia*, A. Jaimes, Ed., ser. ACM Conferences, New York, NY: ACM, 2013, pp. 835–838, ISBN: 9781450324045. https://doi.org/10.1145/2502081.2502224.

17. J. S. Gomez-Canon, E. Cano, T. Eerola, *et al.*, "Music emotion recognition: Toward new, robust standards in personalized and context-sensitive applications," *IEEE Signal Processing Magazine*, vol. 38, no. 6, pp. 106–114, 2021, ISSN: 1053-5888. https://doi.org/10.1109/MSP.2021.3106232.

18. J. S. Gómez-Cañón, E. Cano, P. Herrera, and E. Gómez, Eds., *Joyful for you and tender for us: the influence of individual characteristics and language on emotion labeling and classification: Zenodo*, 2020. https://doi.org/10.5281/ZENODO.4076720.

19. I. Goodfellow, A. Courville, and Y. Bengio, *Deep learning* (Adaptive computation and machine learning). Cambridge, Massachusetts: The MIT Press, 2016, ISBN: 0262337371.

20. T. Greer, X. Shi, B. Ma, and S. Narayanan, *Multi-modal, Multi-task, Music BERT: A Context-Aware Music Encoder Based on Transformers*. 2022. https://doi.org/10.21203/rs.3.rs-2090671/v1.

21. J. Grekow, *From Content-based Music Emotion Recognition to Emotion Maps of Musical Pieces* (Springer eBook Collection Engineering). Cham: Springer, 2018, vol. 747, ISBN: 9783319706092. https://doi.org/10.1007/978-3-319-70609-2.

22. D. Han, Y. Kong, J. Han, and G. Wang, "A survey of music emotion recognition," *Frontiers of Computer Science*, vol. 16, no. 6, 2022, ISSN: 2095-2228. https://doi.org/10.1007/s11704-021-0569-4.

23. T. Hastie, R. Tibshirani, and J. H. Friedman, *The elements of statistical learning: Data mining, inference, and prediction* (Springer series in statistics), 2. ed., corr. at 4. print. New York, NY: Springer, 2009, ISBN: 978-0-387-84858-7.

24. N. He and S. Ferguson, "Music emotion recognition based on segment-level two-stage learning," *International Journal of Multimedia Information Retrieval*, vol. 11, no. 3, pp. 383–394, 2022, ISSN: 2192-662X. https://doi.org/10.1007/s13735-022-00230-z.

25. K. Hevner, "Experimental studies of the elements of expression in music," *The American Journal of Psychology*, vol. 48, no. 2, p. 246, 1936, ISSN: 00029556. https://doi.org/10.2307/1415746.

26. S. Hizlisoy, S. Yildirim, and Z. Tufekci, "Music emotion recognition using convolutional long short term memory deep neural networks," *Engineering Science and Technology, an International Journal*, vol. 24, no. 3, pp. 760–767, 2021, ISSN: 2215-0986. https://doi.org/10.1016/j.jestch.2020.10.009.

27. S. Hochreiter and J. Schmidhuber, "Long short-term memory," *Neural computation*, vol. 9, no. 8, pp. 1735–1780, 1997, ISSN: 0899-7667. https://doi.org/10.1162/neco.1997.9.8.1735.

28. Huaping Liu, Yong Fang, and Qinghua Huang, "Music emotion recognition using a variant of recurrent neural network," *Proceedings of the 2018 International Conference on Mathematics, Modeling, Simulation and Statistics Application (MMSSA 2018)*, 2019.

29. Jacopo de Berardinis, A. Cangelosi, and E. Coutinho, "The multiple voices of musical emotions: Source separation for improving music emotion recognition models and their interpretability," 2020.

30. N. Ketkar and J. Moolayil, *Deep learning with Python: Learn best practices of deep learning models with Pytorch*, Second edition. New York, NY: Apress, 2021, ISBN: 9781484253649.

31. D. P. Kingma and J. Ba, *Adam: A method for stochastic optimization*, 2014.

32. M. Kuhn and K. Johnson, *Applied predictive modeling*, Corrected at 5th printing. New York: Springer, 2016, ISBN: 9781461468493.

33. Y. LeCun, Y. Bengio, and G. Hinton, "Deep learning," *Nature*, vol. 521, no. 7553, pp. 436–444, 2015. https://doi.org/10.1038/nature14539.

34. Y. Li, R. Yuan, G. Zhang, *et al.*, *Mert: Acoustic music understanding model with large-scale self-supervised training*.

35. Y. Li, R. Yuan, G. Zhang, *et al.*, *Map-music2vec: A simple and effective baseline for self-supervised music audio representation learning*.

36. R. Liaw, E. Liang, R. Nishihara, P. Moritz, J. E. Gonzalez, and I. Stoica, *Tune: A research platform for distributed model selection and training*.

37. Z. C. Lipton, J. Berkowitz, and C. Elkan, *A critical review of recurrent neural networks for sequence learning*.

38. T. M. Mitchell, *Machine learning* (McGraw-Hill series in computer science). New York, NY u.a.: McGraw-Hill, 1997, ISBN: 9780071154673.

39. M. Müller, *Fundamentals of music processing: Using Python and Jupyter notebooks*, Second edition. Cham, Switzerland: Springer, 2021, ISBN: 9783030698072.

40. Nitish Srivastava, Geoffrey Hinton, Alex Krizhevsky, Ilya Sutskever, and Ruslan Salakhutdinov, "Dropout: A simple way to prevent neural networks from overfitting," *Journal of Machine Learning Research*, vol. 15, no. 56, pp. 1929–1958, 2014, ISSN: 1533-7928.

41. R. Orjesek, R. Jarina, and M. Chmulik, "End-to-end music emotion variation detection using iteratively reconstructed deep features," *Multimedia Tools and Applications*, vol. 81, no. 4, pp. 5017–5031, 2022, ISSN: 1573-7721. https://doi.org/10.1007/s11042-021-11584-7.

42. R. Panda, R. Malheiro, and R. P. Paiva, "Novel audio features for music emotion recognition," *IEEE Transactions on Affective Computing*, vol. 11, no. 4, pp. 614–626, 2020, ISSN: 1949-3045. https://doi.org/10.1109/TAFFC.2018.2820691.

43. R. Panda, R. M. Malheiro, and R. P. Paiva, "Audio features for music emotion recognition: A survey," *IEEE Transactions on Affective Computing*, p. 1, 2020, ISSN: 1949-3045. https://doi.org/10.1109/TAFFC.2020.3032373.

44. J. Posner, J. A. Russell, and B. S. Peterson, "The circumplex model of affect: An integrative approach to affective neuroscience, cognitive development, and psychopathology," *Development and psychopathology*, vol. 17, no. 3, pp. 715–734, 2005, ISSN: 0954-5794. https://doi.org/10.1017/S0954579405050340.

45. J. A. Russell, "A circumplex model of affect," *Journal of Personality and Social Psychology*, vol. 39, no. 6, pp. 1161–1178, 1980, ISSN: 0022-3514. https://doi.org/10.1037/h0077714.

46. H. Sak, A. Senior, and F. Beaufays, *Long short-term memory based recurrent neural network architectures for large vocabulary speech recognition*, 2014.

47. A. S. Sams and A. Zahra, "Multimodal music emotion recognition in indonesian songs based on cnn-lstm, xlnet transformers," *Bulletin of Electrical Engineering and Informatics*, vol. 12, no. 1, pp. 355–364, 2022, ISSN: 2089-3191. https://doi.org/10.11591/eei.v12i1.4231.

48. D. Schmidtke, T. Schröder, A. Jacobs, and M. Conrad, "Angst: Affective norms for german sentiment terms derived from the affective norms for english words," *Behavior research methods*, vol. 46, Jan. 2014. https://doi.org/10.3758/s13428-013-0426-y.

49. M. Schuster and K. K. Paliwal, "Bidirectional recurrent neural networks," *IEEE Transactions on Signal Processing*, vol. 45, no. 11, pp. 2673–2681, 1997, ISSN: 1941-0476. https://doi.org/10.1109/78.650093.

50. M. Soleymani, M. N. Caro, E. M. Schmidt, C.-Y. Sha, and Y.-H. Yang, "1000 songs for emotional analysis of music," in *Proceedings of the 2Nd ACM International Workshop on Crowdsourcing for Multimedia*, vol. CrowdMM '13, ACM, 2013, pp. 1–6, ISBN: 978-1-4503-2396-3. https://doi.org/10.1145/2506364.2506365.

51. L. Tunstall, L. von Werra, and T. Wolf, *Natural language processing with transformers: Building language applications with Hugging Face*, Revised Edition. Beijing et al.: O'Reilly, 2022, ISBN: 9781098136765.

52. A. Vaswani, N. Shazeer, N. Parmar, *et al.*, *Attention is all you need*.

53. Y.-H. Yang, *Music Emotion Recognition* (Multimedia Computing, Communication and Intelligence). Boca Raton: Taylor & Francis Group, 2011, ISBN: 9781439850473.

54. Y.-H. Yang, Y.-C. Lin, and H. Chen, "Personalized music emotion recognition," in *Proceedings of the 32nd international ACM SIGIR conference on Research and development in information retrieval*, J. Allan, Ed., ser. ACM Conferences, New York, NY: ACM, 2009, pp. 748–749, ISBN: 9781605584836. https://doi.org/10.1145/1571941.1572109.

55. Y.-H. Yang, Y.-C. Lin, Y.-F. Su, and H. H. Chen, "A regression approach to music emotion recognition," *IEEE Transactions on Audio, Speech, and Language Processing*, vol. 16, no. 2, pp. 448–457, 2008, ISSN: 1558-7916. https://doi.org/10.1109/TASL.2007.911513.

56. X. Yang, Y. Dong, and J. Li, "Review of data features-based music emotion recognition methods," *Multimedia Systems*, vol. 24, no. 4, pp. 365–389, 2018, ISSN: 1432-1882. https://doi.org/10.1007/s00530-017-0559-4.

57. Y.-Y. Yang, M. Hira, Z. Ni, *et al.*, *Torchaudio: Building blocks for audio and speech processing*, 2021.

58. Yann LeCun, "Generalization and network design strategies," *Connectionism in perspective*, 1989.

59. Ye Ma, Xinxing Li, Mingxing Xu, Jia Jia, and Lianhong Cai, "Multi-scale context based attention for dynamic music emotion prediction," *Proceedings of the 25th ACM international conference on Multimedia*, 2017.

60. M. Zhang, Y. Zhu, W. Zhang, Y. Zhu, and T. Feng, "Modularized composite attention network for continuous music emotion recognition," *Multimedia Tools and Applications*, vol. 82, no. 5, pp. 7319–7341, 2023, ISSN: 1573-7721. https://doi.org/10.1007/s11042-022-13577-6.

GPSR Compliance
The European Union's (EU) General Product Safety Regulation (GPSR) is a set
of rules that requires consumer products to be safe and our obligations to
ensure this.

If you have any concerns about our products, you can contact us on

ProductSafety@springernature.com

In case Publisher is established outside the EU, the EU authorized
representative is:

Springer Nature Customer Service Center GmbH
Europaplatz 3
69115 Heidelberg, Germany

www.ingramcontent.com/pod-product-compliance
Ingram Content Group UK Ltd.
Pitfield, Milton Keynes, MK11 3LW, UK
UKHW020255070925
462651UK00001B/2